KT-433-518

NATURE VIA NURTURE

NATURE VIA NURTURE

Genes, experience and what makes us human

Matt Ridley

FOURTH ESTATE • *London*

First published in Great Britain in 2003 by
Fourth Estate
A Division of HarperCollins*Publishers*
77–85 Fulham Palace Road,
London W6 8JB
www.4thestate.com

1 3 5 7 9 10 8 6 4 2

A catalogue record for this book is available from the British Library

ISBN 1–84115–745–7

Typeset by Rowland Phototypesetting Ltd,
Bury St Edmunds, Suffolk

Printed in Great Britain by
Clays Ltd, St Ives plc

For Jim

CONTENTS

PROLOGUE

Twelve hairy men

Perverse Mankind! Whose wills, created free,
Charge all their woes on absolute Decree;
All to the dooming Gods their guilt translate,
And follies are miscall'd the crimes of Fate.

Homer's *Odyssey*, translated by
Alexander Pope[1]

'Revealed: the secret of human behaviour,' read the banner headline in the British Sunday newspaper the *Observer* on 11 February 2001. 'Environment, not genes, key to our acts.' The source of the story was Craig Venter, the self-made man of genes who had built a private company to read the full sequence of the human genome (his own) in competition with an international consortium funded by taxes and charities. That sequence – a string of three billion letters composed in a four-letter alphabet containing the complete recipe for building and running a human body – was to be published later in the week. The first analysis had revealed that there were just 30,000 genes in the human genome, not the 100,000 that many had been estimating up until a few months before.

Details had already been circulated to journalists under embargo.

But Venter spilt the story at an open meeting in Lyons on 9 February. Robin McKie of the *Observer* was in the audience and recognised at once that the 30,000 figure was now public. He went up to Venter and asked him if he realised that this broke the embargo; he did. Not for the first time in the increasingly bitter rivalry over the genome project, Venter's version of the story would hit the headlines before that of his rivals. 'We simply do not have enough genes for this idea of biological determinism to be right,' Venter said to McKie. 'The wonderful diversity of the human species is not hard-wired in our genetic code. Our environments are critical.'[2]

Seeing the *Observer*'s first edition, other newspapers followed suit. 'Genome discovery shocks scientists: genetic blueprint contains far fewer genes than thought – DNA's importance downplayed,' proclaimed the *San Francisco Chronicle* later that Sunday.[3] The scientific journals promptly lifted the embargo and the story was in newspapers around the world. 'Analysis of human genome discovers far fewer genes,' intoned the *New York Times*.[4] Not only had McKie scooped the story; Venter had set the theme.

This was the making of a new myth. In truth, the number of human genes changed nothing. Venter's remarks concealed two massive non sequiturs: first, that fewer genes implied more environmental influences, and second, that 30,000 genes were 'too few' to explain human nature where 100,000 would have been enough. As Sir John Sulston, one the leaders of the Human Genome Project, put it to me a few weeks later, just 33 genes, each coming in just two varieties (such as on or off), would be enough to make every human being in the world unique. There are more than ten billion ways of flipping a coin 33 times. So 30,000 does not look such a small number after all. Two multiplied by itself 30,000 times produces a number larger than the total number of particles in the known universe. Besides, if fewer genes meant more free will, that made fruit flies freer than people, bacteria freer still, and viruses the John Stuart Mills of biology.

Fortunately, there was no need for such sophisticated calculations to reassure the population. People were not seen weeping in the street at the humiliating news that our genome had less than twice as many genes as a worm's. Nothing had been hung on the number 100,000, which was just a bad guess. But it was fitting after a century of increasingly repetitive argument over environment versus heredity that the publication of the human genome should be broken on the Procrustean bed of nature-versus-nurture. It was, with the possible exception of the Irish question, the intellectual argument that had changed least in the century just ended. It had divided fascists from communists as neatly as their politics. It had continued unabashed through the discovery of chromosomes, DNA and Prozac. It was fated to be just as bitterly debated in 2003 as it was in 1953, the year of the discovery of the structure of the gene, or in 1900, the year modern genetics began. Even the human genome, at its birth, was being claimed for nurture-versus-nature.

For more than fifty years sane voices have called for an end to the debate. Nature-versus-nurture has been declared everything from dead and finished to futile and wrong – a false dichotomy. Everybody with an ounce of common sense knows that human beings are a product of a transaction between the two. Yet nobody could stop the argument. Immediately after calling the debate futile or dead, the protagonist would charge into the battle himself and start accusing others of overemphasising one or other extreme. The two sides of this argument are the nativists, who I will sometimes call geneticists, hereditarians or naturians; and the empiricists, who I will sometimes call environmentalists or nurturists.

Let me at once play my cards face up. I believe human behaviour has to be explained by both nature and nurture. I am not backing one side over the other. But that does not mean I am taking a 'middle of the road' compromise. As Jim Hightower, a Texan politician, once said: 'There ain't nothing in the middle of the road but a yellow line and a dead armadillo.' I intend to make the case that the genome has indeed changed everything, not by closing the argument or winning the battle for one side or the other, but by enriching it from both ends till they meet in the middle. The discovery of how genes actually influence human behaviour, and how human behaviour influences genes, is about to recast the debate entirely. No longer is it nature-versus-nurture, but nature-via-nurture. Genes are designed to take their cues from nurture. To appreciate what has happened, you will have to abandon cherished notions and open your mind. You will have to enter a world where your genes are not puppet masters pulling the strings of your behaviour, but are puppets at the mercy of your behaviour; a world where instinct is not the opposite of learning, where environmental influences are sometimes less reversible than genetic ones, and where nature is designed for nurture. These cheap and seemingly empty phrases are coming to life for the first time in science. I intend to tell bizarre stories from the deepest recesses of the genome to show how the human brain is built for nurture. My argument in a nutshell is this: the more we lift the lid on the genome, the more vulnerable to experience genes appear to be.

I imagine a photograph taken in the year 1903. It is of a group of men gathered at some international meeting, in a fashionable spot like Baden-Baden or Biarritz, perhaps. 'Men' is not quite the right word, for though there are no women, there is one little boy, one baby and one ghost; the rest are middle-aged or elderly men, mostly rich and all white. There are 12 of them and, as befits the age, there is a great deal of facial hair. There are two Americans, two Austrians, two Britons, two Germans, one Dutchman, one Frenchman, one Russian and one Swiss.

It is, alas, an imaginary photograph, for most of them never met each other. But, like the famous group photograph of physicists at Solvay in 1927 – the one that includes Einstein and Bohr and Marie Curie and Planck and Schrödinger and Heisenberg and Dirac – my picture would capture that moment of ferment when a scientific endeavour throws up a host of new ideas.[5] My 12 men were the ones who put together the chief theories of human nature that came to dominate the twentieth century.

The ghost that hovers overhead is Charles Darwin, dead for 11 years by the time of the photograph, and with the longest beard of all. Darwin's idea is to seek the character of man in the behaviour of the ape and to demonstrate that there are universal features of human behaviour, like smiling. The elderly gent sitting bolt-upright on the far left is his cousin, Francis Galton, 81 years old but going strong; his whiskers hang down the sides of his face like white mice. Galton is the fervent champion of heredity. Next to him sits the American William James, 61, with a square, untidy beard. He is a champion of instinct and maintains that human beings have more impulses than other animals, not fewer. On Galton's right is a botanist, out of place in a group concerned with human nature, and frowning unhappily behind his straggly beard. He is Hugo De Vries, 55, the Dutchman who discovered the laws of heredity only to realise he had been beaten to them more than 30 years before by a Moravian monk named Gregor Mendel. Beside him is the Russian, Ivan Pavlov, 54, his beard full and grey. He is a champion of empiricism, believing that the key to the human mind lies in the conditioned reflex. At his feet, uniquely clean-shaven, sits John Broadus Watson, who will turn Pavlov's ideas into 'behaviourism' and famously claim to be able to alter personality at will merely by training. To Pavlov's right stands the plump, bespectacled, moustachioed German, Emil Kraepelin, and the neatly bearded Viennese, Sigmund Freud, both 47 and both in the throes of influencing generations of psychiatrists away from 'biological' explanations and towards two very different notions of personal history. Beside him is the pioneer of sociology, the Frenchman Emile Durkheim, 45 and especially bushy in beard, busy insisting on the reality of social facts as more than the sum of their parts. His soulmate in this is standing next to him: a German-American (he had emigrated in 1885), the dashing Franz Boas, 45, with his drooping moustaches and duelling scar and his growing insistence that culture shapes human nature, not the other way round. The little boy in the front is the Swiss Jean Piaget, whose theories of imitation and learning will come to fruition, beardless, in mid-century. The baby in the pram is the Austrian Konrad Lorenz, who will revive the study of instinct and describe the vital concept of imprinting in the 1930s, while growing a fine white goatee.

I am not going to claim that these were necessarily the greatest students of human nature, nor that they were all equally brilliant. There are many, both dead and unborn, who would otherwise deserve inclusion in the photograph. David Hume and Immanuel Kant ought to be there, but they died too long ago (only Darwin manages to cheat death for the occasion); so should the modern theorists George Williams, William Hamilton and Noam Chomsky, but they were unborn. So should Jane Goodall, who discovered individuality in apes. So perhaps should some of the more perceptive novelists and playwrights.

But I am going to claim something rather surprising about these 12 men. They were right. Not right all the time, not even wholly right, and I do not mean morally right. They nearly all went too far in trumpeting their own ideas and criticising each other's. One or two of them deliberately or accidentally give birth to grotesque perversions of 'scientific' policy that will haunt their reputations forever. But they were right in the sense that they all contributed an original idea with a germ of truth in it; they all placed a brick in the wall.

Human nature is indeed a combination of Darwin's universals, Galton's heredity, James's instincts, De Vries's genes, Pavlov's reflexes, Watson's associations, Kraepelin's history, Freud's formative experience, Boas's culture, Durkheim's division of labour, Piaget's development and Lorenz's imprinting. You can find all these things going on in the human mind. No account of human nature would be complete without them all.

But – and here is where I begin to tread new ground – it is entirely misleading to place these phenomena on a spectrum from nature to nurture, from genetic to environmental. Instead, to understand each and every one of them, you need to understand genes. It is genes that allow the human mind to learn, to remember, to imitate, to imprint, to absorb culture and to express instincts. Genes are not puppet masters, nor blueprints. Nor are they just the carriers of heredity. They are active during life; they switch each other on and off; they respond to the environment. They may direct the construction of the body and brain in the womb, but then they set about dismantling and rebuilding what they have made almost at once – in response to experience. They are both cause and consequence of our actions. Somehow the adherents of the 'nurture' side of the argument have scared themselves silly at the power and inevitability of genes, and missed the greatest lesson of all: the genes are on their side.

CHAPTER ONE

The paragon of animals

Is man no more than this? Consider him well: Thou owest the worm no silk, the beast no hide, the sheep no wool, the cat no perfume:—Ha! here's three of us are sophisticated!—Thou art the thing itself: unaccommodated man is no more but such a poor, bare, forked animal as thou art.

King Lear[1]

Similarity is the shadow of difference. Two things are similar by virtue of their difference from another; or different by virtue of one's similarity to a third. So it is with individuals. A short man is different from a tall man, but two men seem similar if contrasted with a woman. So it is with species. A man and a woman may be very different, but by comparison with a chimpanzee, it is their similarities that strike the eye – the hairless skin, the upright stance, the prominent nose. A chimpanzee, in turn, is similar to a human being when contrasted with a dog: the face, the hands, the 32 teeth, and so on. And a dog is like a person to the extent that both are unlike a fish. Difference is the shadow of similarity.

Consider, then, the feelings of a naïve young man, as he stepped ashore in Tierra del Fuego on 18 December 1832 for his first encounter with what we would now call hunter-gatherers, or what he would call 'man in a state of nature'. Better still, let him tell us the story:

> It was without exception the most curious & interesting spectacle I ever beheld. I would not have believed how entire the difference between savage & civilized man is. It is much greater than between a wild & domesticated animal, in as much as in man there is greater power of improvement . . . I believe if the world was searched, no lower grade of man could be found.[2]

The effect on Charles Darwin was all the more shocking, because these were not the first Fuegian natives he had seen. He had shared a ship with three who had been transported to Britain, dressed in frocks and coats and taken to meet the king. To Darwin they were just as human as any other person. Yet here were their relatives, suddenly seeming so much less human. They reminded him of . . . well, of animals. A month later, on finding the camp site of a single Fuegian limpet hunter in an even more remote spot, he wrote in his diary: 'We found the place where he had slept – it positively afforded no more protection than the form of a hare. How very little are the habits of such a being superior to those of an animal.'[3] Suddenly, he is writing not just about difference (between civilised and savage man), but about similarity – the affinity between such a man and an animal. The Fuegian is so different from the Cambridge graduate that he begins to seem similar to an animal.

Six years after his encounter with the Fuegian natives, in the spring of 1838, Darwin visited London Zoo and there for the first time saw a great ape. It was an orang-utan named Jenny, and it was the second ape to be brought to the zoo. Its predecessor, Tommy, a chimpanzee, was exhibited at the zoo for a few months in 1835 before he died of tuberculosis. Jenny was acquired by the zoo in 1837, and like Tommy she caused a small sensation in London society. She seemed such a human animal, or was it such a beastly person? Apes posed uncomfortable questions about the distinction between people and animals, between reason and instinct. Jenny featured on the cover of the *Penny Magazine of the Society for the Diffusion of Useful Knowledge*, where the editorial reassured readers that 'extraordinary as the Orang may be compared with its fellows of the brute creation, still in nothing does it trench upon the moral or mental provinces of man'. Queen Victoria, who saw a different orang-utan at the zoo in 1842, begged to differ. She described it as 'frightful and painfully and disagreeably human'.[4]

After his first encounter with Jenny in 1838, Darwin returned to the zoo twice more a few months later. He came armed with a mouth organ, some peppermint and a sprig of verbena. Jenny seemed to appreciate all three. She seemed 'astonished beyond measure' at her reflection in a mirror. He wrote in his notebook: 'Let man visit Ouranoutang in domestication . . . see its intelligence . . . and then let him boast of his proud pre-eminence . . . Man in his arrogance thinks himself a great work, worthy the interposition of a deity. More humble and I believe true to consider him created from animals.' He was applying to animals what he had been taught to apply to geology: the uniformitarian principle that the forces shaping the landscape today are the same as those that shaped the distant past. Later that September, while reading Malthus's essay on population, he had his sudden insight into what we now know as natural selection.

Jenny had played her part. When she took the mouth organ from him and placed it to her lips, she had helped him realise how high above the brute some animals could rise, just as the Fuegians had made him realise how low beneath civilisation some humans could sink. Was there a gap at all?

He was not the first person to think this way. Indeed, a Scottish judge, Lord Monboddo, had speculated in the 1790s that orang-utans could speak – if educated. Jean-Jacques Rousseau was only one of several Enlightenment philosophers who wondered if apes were not continuous with 'savages'. But it was Darwin who changed the way human beings think of their own nature. Within his lifetime, he saw educated opinion come to accept that human bodies were those of just another ape modified by descent from a common ancestor.

But Darwin had less success in persuading his fellow human beings that the same argument could apply to the mind. His consistent view, from his earliest notebooks after reading David Hume's *Treatise on*

Human Nature to his last book, about earthworms, was that there was similarity, rather than difference, between human and animal behaviour. He tried the same mirror test on his children that he had tried on Jenny. He continually speculated on the animal parallels and evolutionary origins of human emotions, gestures, motives and habits. As he stated plainly, the mind needed evolution as much as the body did.

But in this he was deserted by many of his supporters, the psychologist William James being a notable exception. Alfred Russel Wallace, for example, the co-discoverer of the principle of natural selection, argued that the human mind was too complex to be the product of natural selection. It must instead be a supernatural creation. Wallace's reasoning was both attractive and logical. It was based on similarity and difference again. Wallace was remarkable for his time in being mostly devoid of racial prejudice. He had lived among natives of South America and South-East Asia and he thought of them as equals, morally if not always intellectually. This led him to the belief that all races of humanity had similar mental abilities, which puzzled him because it implied that in most 'primitive' societies, the great part of human intelligence went unused. What was the point of being able to read or do long division if you were going to spend all your life in a tropical jungle? Ergo, said Wallace, 'some higher intelligence directed the process by which the human race was developed'.[5]

We now know that Wallace's assumption was entirely right, where Darwin's was wrong. The gap between the 'lowest' human and the 'highest' ape is enormous. Genealogically, we all descend from a very recent common ancestor who lived just 150,000 years ago, whereas the last common ancestor with a chimpanzee lived at least five million years ago. Genetically, the differences between a human being and a chimpanzee are at least ten times as numerous as those between the two most dissimilar human beings. But Wallace's deduction from this assumption, that therefore the human mind required a different kind of explanation from the animal mind, is not warranted. Just because two animals are different does not mean they cannot also be similar.

René Descartes had decreed firmly in the seventeenth century that people were rational and animals were automata. 'They act not from knowledge but from the disposition of their organs . . . Brutes not only have a smaller degree of reason than men, but are wholly lacking in it.'[6] Darwin dented this Cartesian distinction for a while. Freed at last from the need to think of the human mind as a divine creation, some of Darwin's contemporaries, the 'instinctivists', began to think of humans as automata driven by instinct; others, the 'mentalists', began to credit the animal brain with reason and thought.

The mentalist anthropomorphism reached its apogee in the work of the Victorian psychologist, George Romanes, who eulogised the intelligence of pets, such as dogs that could lift latches and cats that seemed to understand their masters. Romanes believed that the only explanation for their behaviour was conscious choice. He went on to argue that each species of animal had a mind just like the human one, only frozen at a stage equivalent to a child of a certain age. Therefore, a chimpanzee had the mind of a young teenager, while a dog was equivalent to a younger child, and so on.[7]

Ignorance of wild animals sustained this notion. So little was known about the behaviour of apes that it was easy to go on thinking of them as primitive versions of people, rather than sophisticated animals that were brilliantly good at being apes. Especially with the discovery of the seemingly fierce gorilla in 1847, encounters between human beings and wild apes were exclusively brief and violent. When apes were brought to zoos, they had little opportunity to show their repertoire of wild habits, and their keepers seemed to evince more interest in their ability to 'ape' human customs than in what came naturally to them. For instance, from the very first arrival of chimpanzees in Europe, there seems to have been an obsession with serving them tea. The great French naturalist, Georges Leclerc, Comte de Buffon, was one of the first 'scientists' to see a captive chimp in about 1790. What did he find worth remarking? That he watched it 'take a cup and saucer and lay them on the table, put in sugar, pour out its tea, leave it to cool without drinking'.[8] Thomas Bewick, a few years later, reported breathlessly that an ape 'shewn in London some years ago was taught to sit at table, make use of a spoon or fork in eating its victuals'.[9] And when Tommy and Jenny reached London Zoo in the 1830s, they were quickly taught to eat and drink at table for the benefit of a paying audience. The tradition of the chimpanzee tea party was born. By the 1920s it was a daily ritual at London Zoo, the chimps trained both to ape human customs and to break them: 'there was the ever present danger that their table manners would become too polished.'[10] The chimpanzee tea parties at zoos ran for 50 years. In 1956, the Brooke Bond company made the first of many hugely successful television commercials for its tea using a chimps' tea party, and Tetley finally dropped its chimps' tea party advertisements only in 2002. By 1960, human beings still knew more about chimps' ability to learn tea-table manners than they did about how the animals behaved in the wild. No wonder apes were viewed as ridiculous apprentice people.

In psychology, mentalism was soon ridiculed and demolished. The early twentieth-century psychologist Edward Thorndike demonstrated that Romanes's dogs invariably learned their clever tricks by accident. They did not understand how a door latch worked; they simply repeated any action that accidentally enabled them to open the door. In reaction to the credulity of mentalism, psychologists began to make the opposite assumption: that animal behaviour was unconscious, automatic and reflex. The assumption was soon a creed. The radical behaviourists who brushed aside the mentalists in the same decade as the Bolsheviks brushed aside the Mensheviks asserted brusquely that animals did not think, reflect or reason; they just responded to stimuli. It became heresy even to talk about animals having mental states, let alone to attribute human understanding to them. Soon, under Burrhus Skinner, the behaviourists would apply the same logic to human beings. After all, people do not just anthropomorphise animals, accuse toasters of perversity and thunderstorms of fury. They also anthropomorphise other people, crediting them with too much reason and too little habit. Try reasoning with a nicotine addict.

But since nobody took Skinner all that seriously on the subject of people, the behaviourists had unwittingly restored the distinction between the human and the animal mind to exactly where Descartes had placed it. Sociologists and anthropologists, with their emphasis on the peculiarly human attribute called culture, had outlawed all talk of human instinct. By the middle of the twentieth century, it was heresy to speak of animal minds, and heresy to speak of human instincts. Difference, not similarity, was all.

THE SIMIAN SOAP OPERA

That was all to change in 1960, when a young woman virtually untrained in science began to watch chimpanzees on the shores of Lake Tanganyika. As she later wrote:

> How naïve I was. As I had not had an undergraduate science education I didn't realise that animals were not supposed to have personalities, or to think, or to feel emotions or pain . . . Not knowing, I freely made use of all those forbidden terms and concepts in my initial attempts to describe, to the best of my ability, the amazing things I had observed at Gombe.[11]

As a result, Jane Goodall's account of life among the chimps of Gombe reads like a soap opera about the Wars of the Roses written by Jane Austen – all conflict and character. We feel the ambition, the jealousy, the deception and the affection; we distinguish personalities; we sense motives; we cannot help but empathise:

> Gradually, Evered's confidence returned – partly, no doubt, because Figan was by no means always with his brother: Faben was still friendly with Humphrey, and Figan, wisely, steered clear of the powerful male. Moreover, even when the brothers were together, Faben did not *always* help Figan: sometimes he just sat and watched.[12]

Though few realised it until later, Goodall's anthropomorphism had driven a stake through the heart of human exceptionalism. Apes were revealed not as blundering, primitive automata, who were bad at being people, but as beings with social lives as complex and subtle as ours. Either human beings must be more instinctive, or animals must be more conscious than we had previously suspected. The similarities, not the differences, were what caught the attention.

Of course, the news that Goodall had narrowed the Cartesian gap travelled very slowly across the divide between animal and human sciences. Even though the very purpose of Goodall's study, as conceived by her mentor the anthropologist Louis Leakey, was to shed light on the behaviour of ancient human ancestors, anthropologists and sociologists were trained to ignore animal findings as irrelevant. When Desmond Morris spelled out the similarities in his book *The Naked Ape* in 1967, he was generally dismissed as a sensationalist by most students of humankind.

Defining human uniqueness had been a cottage industry for philosophers for centuries. Aristotle said man was a political animal. Descartes said we were the only creature that could reason. Marx said we alone were capable of conscious choice. Now only by heroically narrow definitions of these concepts could Goodall's chimps be excluded.

St Augustine said we were the only creature to have sex for pleasure rather than procreation. (A reformed libertine should know.) Chimpanzees begged to differ, and their southern relatives, bonobos, were soon to blow the definition to smithereens. Bonobos have sex to celebrate a good meal, to end an argument or to cement a friendship. Since much of this sex is homosexual or with juveniles, procreation cannot even be an accidental side effect.

Then we thought we were the only species to make and use tools. One of the first things Jane Goodall observed was chimpanzees fashioning stalks of grass to extract termites, or crushing sponges of leaves to get drinking water. Leakey telegraphed her ecstatically: 'Now we must redefine tool, redefine man, or accept chimpanzees as humans.'

Next we told ourselves that we alone had culture: the ability to transmit acquired habits from one generation to the next by imitation. But what are we to make of the chimpanzees of the Tai forest in West Africa, which for many generations have taught their young to crack nuts using wooden hammers on a rock anvil? Or the killer whales that have utterly different hunting traditions, calling patterns and social systems according to which population they belong to?[13]

We had assumed we were the only animal to wage war and to kill our fellows. But in 1974 the chimps of Gombe (and subsequently most other colonies studied in Africa) put paid to that theory by raiding silently into the territory of neighbouring troops, ambushing the males and beating them to death.

We still believed we were the only animal with language. But then we discovered monkeys have a vocabulary for referring to different predators and birds, while apes and parrots are capable of learning quite large lexicons of symbols. So far there is nothing to suggest that any other animal can acquire a true grasp of grammar and syntax, though the jury is still out for dolphins.

Some scientists believe that chimpanzees do not have a 'theory of mind': that is, they cannot imagine what another chimpanzee is thinking. If so, for example, they could not act upon the knowledge that another individual holds a false belief. But experiments are ambiguous. Chimps regularly engage in deception. In one case, a baby chimp pretended that he was being attacked by an adolescent in order to get his mother to allow him to suckle from her nipple.[14] It certainly looks as if they are capable of imagining how other chimps think.

More recently, the argument that only human beings have subjectivity has been revived. The author Kenan Malik argues that 'humans simply are not like other animals and to assume we are is irrational . . . Animals are objects of natural forces, not potential subjects of their own destiny.' Malik's point is that because we, uniquely, possess consciousness and agency, so we alone can break out of the prison of our heads and go beyond a solipsistic view of the world. Yet I would argue that consciousness and agency are not confined to human beings, any more than instinct is confined to non-human animals. See almost any passage of Goodall's books for evidence. Even baboons have recently performed well enough at computer discrimination tasks to show they are capable of abstract reasoning.[15]

This debate has been running for more than a century. In 1871 Darwin drew up a list of human peculiarities that had been claimed to form an impassable barrier between man and animals. He then demolished each peculiarity one by one. Though he believed only man had a fully developed moral sense, he devoted a whole chapter to the argument that a moral sense was present, in primitive form, in other animals. His conclusion was stark:

> The difference in mind between man and the higher animals, great as it is, is certainly one of degree and not of kind. We have seen that the senses and intuitions, the various emotions and faculties, such as love, memory, attention, curiosity, imitation, reason, &c., of which man boasts, may be found in an incipient, or even sometimes in a well-developed condition, in the lower animals.[16]

Wherever you look there are similarities between our behaviour and that of animals, which cannot be simply swept under the Cartesian carpet. Yet, of course, it would be perverse to argue that people are no different from apes. The truth is we are different. We are more capable of self-awareness, of calculation and of altering our surroundings than any other animal. Clearly, in some sense, this sets us apart. We have built cities, travelled in space, worshipped gods and written poetry. Each of these things owes something to our animal instincts – shelter, adventure and love – but that rather misses the point of them. It is when we go beyond instinct that we seem most idiosyncratically human. Perhaps, as Darwin suggested, the difference is one of degree rather than kind; it is quantitative, not qualitative. We can count better than chimpanzees; we can reason better, think better, communicate better, emote better, perhaps even worship better. Our dreams are probably more vivid, our laughter more intense, our empathy more profound.

Yet that leads straight back to mentalism, equating an ape with an apprentice person. Modern mentalists have diligently tried to teach animals to 'speak'. Washoe (a chimp), Koko (a gorilla), Kanzi (a bonobo) and Alex (a parrot) have all done remarkably well. They have learned hundreds of words, usually in the form of sign language, and have learned to combine them into primitive phrases. Yet, as Herbert

Terrace pointed out after doing the same with a chimpanzee called Nim Chimpsky, all these experiments have taught us is how bad these animals are at language. They rarely even rival a two-year-old child, and they seem incapable of using syntax and grammar except by accident. As Stalin is reputed to have said of military force, quantity has a quality all its own. We are so much better at language than even the cleverest ape that it really could be called a difference of kind, not degree. That is not to say it does not have roots and homologies in animal communication, but then a bat's wing has homology with a frog's front foot, and a frog cannot fly. To concede that language is a qualitative difference does not imply that we can set human beings apart from nature, though. Trunks are unique to elephants. Spitting venom is unique to cobras. Uniqueness is not unique.

So which are we, similar to apes or different from apes? Both. The argument about human exceptionalism, today as in Victorian times, is mired in a simple confusion. People still insist that their opponents must take sides: either we are instinctive animals, or we are conscious beings, but we cannot be both. Yet both similarity and difference can be true at the same time. You do not have to abandon an ounce of human agency when you accept the kinship of our minds with those of apes.[17] Neither similarity nor difference wins; they coexist. Let some scientists study the similarities while others study the differences. It is time we abandoned what the philosopher Mary Midgley has called 'the strange segregation of humans from their kindred that has deformed much of enlightenment thought'.[18]

SEX AND ITS EFFECTS

There is one way in which behaviour seems to evolve differently from anatomy. In the case of anatomy, most similarities are the result of common descent, or what evolutionists call phylogenetic inertia. For example, human beings and chimpanzees both have five digits on each hand and foot. This is not because five is the perfect number for the lifestyle of both species, but because among the early amphibians, one happened to have five digits and most of its myriad descendants, from frogs to bats, have not altered the basic pattern. Some, like birds and horses, do have fewer digits, but none of the apes do.

The same is not true of social behaviour. By and large, ethologists have found very little phylogenetic inertia in social systems. Closely related species can have very different social organisation if they live in different habitats or eat different food. Distant relatives can have very similar social systems by convergent evolution if they inhabit similar ecological niches. Where two species show similar behaviour, it tells you less about their common ancestor and more about the pressures of the environment that shaped them.[19]

A good example is the sex life of the African apes. As primatologists delved further into the lives of apes, they found that alongside the similarities were some intriguing contrasts. These contrasts were thrown into sharper relief by the studies of George Schaller and Diane Fossey on gorillas, Birute Galdikas on orang-utans and the later studies of Takayoshi Kano on bonobos. In the zoo, a chimp looks a bit like a small gorilla. The skeletons of large chimpanzees have been confused with those of small gorillas. In the wild, however, there is a marked difference in their behaviour. It all starts with diet. Gorillas are herbivores, eating the stems and leaves of green plants such as nettles or reeds as well as some fruit. Chimpanzees are principally frugivores, seeking out fruit in trees, but adding ants, termites or monkey meat when they can. This difference in diet dictates a difference in social organisation. Plants are abundant but not very nutritious. To thrive on them, a gorilla must spend nearly all day eating and need not move very far. This makes a group of gorillas rather stable and easy to defend. This in turn has tempted male gorillas into evolving a polygamous mating strategy: each male can monopolise a small harem of females and their immature young, driving away other males.

Fruit, however, appears unpredictably in different places. Chimpanzees need to have large home ranges to be sure of finding a fruiting tree. But when a tree is found there is plenty of food to go round, so the animals can share their home range with many other chimps. But because of the large home range, these groups often split up temporarily. Consequently, for the male chimp, the polygamy strategy does not work. The only way to control access to such a large group of females is to share the job with other males. Hence the sexual favours of a troop of chimps are shared among an alliance of males. One becomes the 'alpha' male and takes a greater share of the matings, but he does not monopolise.

This difference in social behaviour, stemming from a difference in diet, was wholly unsuspected until the 1960s. And it was only in the 1980s that a remarkable consequence became clear. The difference has left its mark on the anatomy of the two ape species. For gorillas the reproductive rewards of owning a harem of females are so great that males which take great risks to get them have generally proved more fecund ancestors than males of a more cautious disposition. And one risk that is worth running is growing to a very large size – even though it takes a lot of food to run a big body. Consequently, an adult male gorilla weighs about twice as much as a female.

Among chimpanzees, males are not under such pressure to be big. For a start, being too big makes it harder to climb trees, and it means you have to spend more time eating. Better to be only a little larger than a female and to use cunning as well as strength to rise to the top of the hierarchy. Besides, there is no point in trying to suppress all sexual rivals, because you will sometimes need them as allies to defend the home range. However, because most females are mating with lots of males within the troop, the male chimps that most often became ancestors were in the past the ones that ejaculated often and voluminously. The competition between male chimps continues inside the female vagina in the form of sperm competition. Consequently, male chimpanzees have gigantic testicles and prodigious sexual stamina. As a proportion of body weight, chimpanzee testicles are 16 times greater than gorilla testicles. And a male chimp has sex approximately one hundred times as often as a male gorilla.

There is a further consequence. Infanticide is common in gorillas as it is in many primates. A bachelor male infiltrates a harem, grabs a baby and kills it. This has two effects on the baby's mother (apart from causing her great, though transient, distress): first, by halting her lactation it brings her back into oestrus; second, it persuades her that she needs a new harem master who is better at protecting her babies. And who better to choose than the raider? So she leaves her mate and marries her baby's killer. Infanticide brings genetic rewards to males, who thereby become more fecund ancestors than males that do not kill babies; hence most modern gorillas are descended from killers. Infanticide is a natural instinct in male gorillas.

But in chimps females have 'invented' a counter-strategy that largely averts infanticide: they share their sexual favours widely. The result is that any ambitious male, if he were to start his reign with a killing spree, might be killing some of his own babies. Males that hold back from baby-killing therefore leave more offspring behind. To confuse paternity by seducing many males into possible fatherhood, the females have evolved exaggerated sexual swellings on their pink bottoms to advertise their fertile periods.[20]

The testicle size of a chimp is a meaningless statistic on its own. It only makes sense by comparison with the gorilla testicles. That is the essence of the science of comparative anatomy. And having looked at two species of African ape in such a way, why not include a third? Anthropologists are fond of claiming an almost limitless diversity of behaviours in human cultures, but there is no human culture so extreme that it even begins to compare with the social system of either the chimpanzee or the gorilla. Not even the most polygamous human society is exclusively organised into harems that are passed from one male to another. Human harems are built up one by one, so that most males, even in societies that encourage polygamy, only have one wife. Likewise, despite various attempts to invent free-love communes, nobody has succeeded in achieving, let alone sustaining, a society in which every man has repeated brief affairs with every woman. The truth is, the human species has just as characteristic a mating system as any other: characterised by long pair bonds, usually monogamous, but occasionally polygamous, embedded in a large chimp-like troop or tribe. Likewise, however variable testicle size is among men, there is no man living whose testicles (as a proportion of body weight) are as small as a gorilla's or as big as a chimpanzee's. As a proportion of body weight, our testicles are nearly five times as large as gorillas' and one-third the size of chimpanzees'. This is compatible with a monogamous species showing a degree of female infidelity. The difference between species is the shadow of the similarity within the species.

An intriguing explanation of the human pair bond once again focuses on food. The primatologist Richard Wrangham puts it down to cooking. With the taming of fire and its adoption for cooking – which is a form of predigestion of food – there came a reduced need for chewing. Suggestive evidence for the controlled use of fire now goes back to 1.6 million years ago, but circumstantial evidence hints that it may have happened even earlier. At around 1.9 million years ago the teeth of human ancestors shrank at the same time as the body size of females grew. This indicates a better diet more easily digested, which in turn sounds like cooking. But cooking requires you to gather food and bring it to the hearth, which would have provided ample opportunities for bullies to steal the fruits of others' labour. Or, since males were at that time much bigger and stronger than females, for males to steal food from females. Accordingly, any female strategy that prevented such theft would have been selected, and the obvious one was for a single female to form a relationship with a single male to help her guard the food they both gathered. These increasingly monogamous males would then not be competing with each other so fiercely for every mating opportunity, which would result in their becoming smaller relative to females – and the sex difference in size began to shrink 1.9 million years ago.[21] Later, the pair bond developed into something even deeper when ancestral human beings invented a sexual division of labour. Among all hunter-gatherers, men are usually more interested in and better at hunting; women are more interested in and better at gathering. The result is an ecological niche that combines the best of both worlds – the protein of meat and the reliability of plant food.[22]

But, of course, there are not three species of African ape; there are four. The bonobos that live to the south of the Congo river may look rather like chimpanzees, but they have been evolving apart for two million years, ever since the river split their ancestral range in two. Like chimps they eat fruit; like chimps they live in large home ranges shared by multi-male troops. It follows that their sex lives, and their testicle size, should be like those of chimpanzees. But, as if to teach us scientific humility, they are astonishingly different. In bonobos, females are usually able to dominate and intimidate males. They do this by forming coalitions and coming to each other's aid. A male bonobo in trouble can count on his mother's support more than he can count on that of his male friends. An adult female bonobo, supported by her best friends, can usually outrank any male.[23]

But why? The secret of the bonobo sisterhood lies in sex. The bond between two female best friends is cemented by frequent and intense bouts of 'hoka-hoka', which scientists unromantically translate into genito-genital rubbing. Under the benign rule of cooperative and loving sisterhoods, the society of the bonobo reads more like a feminist fantasy than something real. That it should come to be understood only in the 1980s, when male-biased science was under challenge, is an uncanny coincidence. (The mind boggles at how the Victorians would have described hoka-hoka.)

As predicted by feminist doctrine, male bonobos have reacted to the new female-dominated regime by evolving kinder, gentler natures. There is much less fighting and shouting, and so far murderous raids on members of other troops are unknown. Since female bonobos are even more sexually active than chimps and have sex nearly ten times as often (and a thousand times as often as gorillas), the ambitious male bonobo's best strategy for attaining fatherhood is to save his energy for the bedchamber, not the boxing ring. I would like to be able to tell you that bonobo testicles are even bigger than chimpanzee testicles, but – although they are certainly very large – nobody has yet managed to weigh them.[24] In her book *Sexual Selections*, Marlene Zuk describes how the timely discovery of bonobo sex lives has made them into the latest animal celebrities, supplanting the dolphins which had rather blotted their eco-friendly image by indulging in something that looks very like kidnapping and gang rape. Inevitably, sex therapists have begun trumpeting the 'bonobo way' of sex. Dr Susan Block (of the Dr

Susan Block Institute for the Erotic Arts and Sciences in Beverly Hills) proclaims that these 'horniest apes on earth' are models for us all if we are to live in peace. 'Liberate your inner bonobo,' she urges. 'You can't very well fight a war while you're having an orgasm.' She pledges a share of the profits from her 'ethical hedonism' television and Internet shows to bonobo conservation.[25]

These are just our closest cousins. The apes of Asia – orang-utans and gibbons – have entirely different sex lives again. So do the many and various species of monkey, presenting a bewildering variety of social and sexual stratagems, each one suited to its habitat and food. Forty years of field primatology have confirmed that we are a unique species, completely unlike any other. There is no exact parallel to the human scheme. But in the animal kingdom, there is nothing exceptional in being unique. Every species is unique.

ENTER GENETICS

The argument about human exceptionalism, swaying between Darwinian similarity and Cartesian difference, shows no sign of ending. Each generation is doomed to fight the same old battles. If you arrive in the world in a time when people have strayed a bit far into anthropomorphic similarity, then you can find a fresh argument for how different animals and people are. If the air is full of difference, then you can champion the similarities. Philosophy is like this: eternally unsettled and only occasionally disturbed by new facts.

Then came an unexpected threat to this pleasant debate. A threat of a resolution. A threat of defining once and for all, at root, what the difference is between a person and a chimpanzee; what you would have to do to a chimpanzee to make it into a person.

It happened about the same time that Jane Goodall was undermining the exceptionalism of human behaviour. Almost completely forgotten until rediscovered in the 1960s was an extraordinary experiment done by a Californian named George Nuttall in 1901 while at

Cambridge University. He noticed that the more closely related two species were, the more their blood produced the same immune reaction in a rabbit. He injected blood from, say, a monkey into a rabbit repeatedly for some weeks, then a few days after the last injection extracted serum from the rabbit's blood. That serum, mixed with the blood of a monkey, caused it to thicken as the immune reaction set in. Mixed with the blood of a different animal, it thickened more according to how closely related the species were. By this means Nuttall established that human beings were more closely related to apes than they were to monkeys. This ought to have been obvious from the lack of a tail and other features, but it was still controversial at the time.

In 1967 at Berkeley, Vincent Sarich and Allan Wilson revived Nuttall's biochemical techniques in a more sophisticated form and used them to construct a 'molecular clock' that measured the actual length of time since two species had shared a common ancestor. They concluded that human beings had shared a common ancestor with the great apes not 16 million years ago, as was then conventional wisdom, but only about five million years ago. Anthropologists, whose fossils implied a more ancient split, reacted with contempt. Sarich and Wilson stuck to their guns. In 1975, Wilson asked his student Marie-Claire King to repeat the exercise for DNA in order to find the genetic differences between human beings and apes. She came back disappointed. It was impossible to find differences, she said, because human and chimpanzee DNA was so astonishingly similar: close to 99 per cent of the DNA in a human being was identical to that in a chimpanzee. Wilson was thrilled: the similarity was more exciting than the difference.

That figure has meandered a little since the 1970s. Most estimates place it at 98.5 per cent, although two recent detailed studies of actual stretches of genome came to a figure of 98.76 per cent.[26] However, just as the 98.5 per cent was seeping into the public consciousness, Roy Britten wrote a dramatic paper in 2002 showing that it was out by a mile. He confirmed that if you count only substitutions – i.e., letters in the text that are different between human and chimpanzee genes –

you do indeed get a figure of 98.6 per cent. But if you then add in the textual insertions or deletions, the figure drops to 95 per cent.[27]

Whatever. It was still a terrible shock to science to discover just how small was the genetic distance between the two species. 'The molecular similarity between chimpanzees and humans is extraordinary because they differ far more than many other [closely related] species in anatomy and way of life,' wrote King and Wilson.[28] An even greater shock was in store in 1984, when Charles Sibley and Jon Ahlquist at Yale found that chimpanzee DNA was more like human DNA than it was like gorilla DNA.[29] This was a moment of human dethronement similar to Copernicus placing the Earth within the solar system as just another planet. Sibley and Ahlquist placed the human species within the ape family as just another ape. From having our own distinct ape lineage stretching back 16 million years, we were now forced to admit that not only did we share a common ancestor not much more than five million years ago, but that we were the most recent branch of the family. Our common ancestor with the chimp lived after the common ancestor of both with the gorilla and long after the common ancestor of all three with the organ-utan. Incredible as it may seem, chimpanzees are more closely related to human beings than they are to gorillas (a conclusion that Britten's reanalysis of the precise number does not alter). Nothing in the anatomy or fossil record of the African apes suggested such a possibility. Human beings are not the odd ones out.

Time has dulled these shocks. But there are more coming. Reading the DNA of a human being, alongside that of a chimpanzee, might once and for all define the difference between them. At the time of writing, the complete genome of the chimpanzee is not yet available. Even when it is, it may prove tricky to work out which differences are the ones that matter. The human genome contains about three billion 'letters' of code. Strictly speaking, these are chemical bases on a molecule of DNA, but since it is their order, not their individual properties, that determines what they produce, they can be treated as digital information. The difference between two individual human beings amounts, on average, to 0.1 per cent, so there are three million different letters between me and my neighbour. The difference between a human being and a chimpanzee is about 15 times as great, or 1.5 per cent. That equates to 45 million different letters. That is about ten times as many letters as there are in the whole Bible, or 75 books the length of this one. The book of digital differences between our two species, unannotated, would fill eleven feet of bookshelf. (The bookshelf of similarities, by contrast, would stretch to 250 yards.)

Look at it another way. Scientists now reckon that there are about 30,000 human genes. That is, scattered throughout the genome are 30,000 distinct stretches of digital information that are directly translated into protein machinery to run and build the body: a gene being a recipe for a protein. Chimpanzees almost certainly have roughly the same number of genes. Since 1.5 per cent of 30,000 is 450, then it seems to follow that we have 450 different, uniquely human genes. Not such a big number. The other 29,550 genes are identical in us and chimps. But this is actually most unlikely. It could instead be that every single human gene is different from every single chimp gene, but only 1.5 per cent of its text is different. The truth is bound to lie somewhere between the two. Many genes will be identical in closely related species; many will be slightly different. A very few will be utterly different.

The most visible difference is that all apes have one more pair of chromosomes than people do. The reason is simple enough to find: at some point in the past, two middle-sized ape chromosomes fused together in the ancestors of all human beings to form the large human chromosome known as chromosome 2. This is a surprising rearrangement, and it almost certainly means that chimp–human hybrids would be sterile if they could survive at all. It may have helped create what evolutionists delicately call 'reproductive isolation' between the species in the past.

But the rearrangement of the chromosomes does not necessarily imply a difference in genetic text at that spot. Although the chimpanzee genome is still largely terra incognita, already there are significant textual differences known between human and chimp (or other ape) genes. For example, whereas people have a mixture of A, B and O blood groups, chimpanzees have only A and O, while gorillas have only B. Likewise, there are three common variants of a human gene called APOE, and chimpanzees only have one – the one most associated with Alzheimer's disease in people. There seems to be a distinct difference in the way thyroid hormones work in people compared with other apes. The significance of this is unknown. And a family of genes on chromosome 16 has undergone several bursts of duplication in the apes after they had separated from the monkey lineage 25 million years ago. Each set of these so-called 'morpheus' genes in human beings has diverged rapidly in sequence from each other and from those in other apes – evolving at nearly 20 times the normal rate. Some of these morpheus genes might indeed be described as uniquely human genes. But exactly what these genes do, or why they are evolving apart so rapidly in apes, remains mysterious.[30]

Most of these differences are also variable among people; there is nothing here unique to human beings as a whole. In the mid-1990s, however, the first genetically unique feature universal to all people and absent from all apes was discovered. Several years before, a medical professor in San Diego named Ajit Varki became intrigued by a unique form of human allergy: an allergy to a particular kind of sugar (a certain 'sialic acid') found attached to proteins in animal serum. This immune reaction is partly responsible for the severe reaction that people often have to horse serum used as a snake-bite antidote, for example. We human beings simply cannot tolerate this 'Gc' version of sialic acid, because we do not have it in the human body. Varki, together with Elaine Muchmore, soon discovered the cause by first noting that unlike human beings, chimpanzees and other great apes did have Gc. The human body does not manufacture Gc sialic acid because it lacks the enzyme for making it from Ac sialic acid. Without the enzyme, human beings cannot add an oxygen atom to the Ac form. All human beings lack the enzyme, but all apes have it. This was the first universally true biochemical difference between us and them. Fittingly, at the end of a millennium that saw us humiliatingly demoted from the centre of the universe and the apple of God's eye to just another ape, Varki now seemed to suggest that we differ by just a single atom on a humble sugar molecule: and an omission at that! Not a promising locus for the soul.

By 1998 Varki knew why we were peculiar: a 92-letter sequence was missing from a gene called CMAH on chromosome 6 in human beings, a gene that codes for the enzyme that makes Gc. Next he discovered how it had gone missing. Right in the middle of the gene was an Alu sequence, a sort of 'jumping gene' of a kind that infests our genome. In the ape genome there is a different and more ancient Alu, but the one in the human gene was of a sequence known to be unique to human beings.[31] So some time after the divergence of the human and chimp lineage, this Alu had done what it does best, which is to jump into the CMAH gene, swap places with the older Alu and accidentally remove the 92-letter chunk of the gene while it was about it. (If this all sounds like double genetic Dutch, try thinking of it this way: a computer virus has destroyed one of your files.)

Varki's discovery initially raised a big yawn from the scientific establishment. So what, they cried, you have found a gene that is bust in human beings but not in apes. Big deal. Varki is not easily discouraged, and by now he was interested by the whole subject of human–ape difference. The first issue was to pinpoint when the mutation had occurred. DNA cannot be recovered from ancient fossils of human ancestors, but sialic acid can be. He found that Neanderthals were like us, in having Ac, but no Gc, but older fossils (from Java and Kenya) were all from warmer climates and their sialic acids had degraded too far. However, by counting the number of changes in the defunct human CMAH gene, and using a molecular clock, his colleague Yuki Takahata has been able to estimate that the change happened about 2.5 or 3 million years ago in some human being who is now one of the ancestors of all people alive.

Varki began to investigate other possible consequences of the mutation. Most other animals seemed to have the working gene, even sea urchins, but if the gene is 'knocked out' in the embryo of a mouse, the mouse grows up healthy and fertile. Sialic acid is a sugar found on the outside of cells, like a sort of flower growing from the cell surface. It is one of the first targets for infectious pathogens including botulism, malaria, influenza and cholera. Lacking one of the common forms of sialic acid might make us more or less vulnerable to these diseases than our ape relatives (cell-surface sugars seem to be a sort of first line of defence in the immune system). But the most intriguing thing about the Gc form of sialic acid is that it is easily found throughout the body of mammals except in the brain. Varki's gene is almost entirely switched off in the brains of mammals. There must be some reason why you cannot operate a mammalian brain properly unless you switch this gene off almost completely. Perhaps, muses Varki, the expansion of the human brain, which accelerated about two million years ago, was made possible by going one further and switching the gene off altogether throughout the body. He admits it is a 'wild idea' for which he has no evidence; he is in uncharted territory. Intriguingly, he has since found another gene concerned with processing sialic acid that is also knocked out in human beings.[32]

Even esoteric research like this may have practical consequences. It gives a strong reason to abandon the idea of xeno-transplantation, the transplanting of animal organs into people: allergic reactions to the Gc sugars in animal organs are almost inevitable. Since you can find traces of Gc sialic acid in human tissues, presumably from animal food, Varki has been drinking diluted Gc sialic acid recently to test how his own body handles it. He wonders if some of the diseases that are caused by eating 'red meat' may be associated with encountering this animal version of the sugar. But Varki is the first to admit that the vast range of differences between human beings and apes cannot be boiled down to one kind of sugar molecule.

We use roughly the same set of genes as other mammals, but we achieve different results with them. How can this be? If two sets of near-identical genes can produce such different-looking animals as a human being and a chimpanzee, then it seems superficially obvious that the source of the difference must lie elsewhere than in the genes. Nurtured as we are in nature–nurture dichotomies, the obvious alternative that occurs to us is nurture. Well, then, do the obvious experiment. Implant a fertilised human egg into the womb of an ape, and vice versa. If nurture is responsible for the difference, the human will give birth to a human and the chimp to a chimp. Any volunteers?

It has been done, though not in apes. In zoos, surrogate mothers have been made to lend their wombs to foetuses from other species in the cause of conservation. The results have been mixed at best. Wild oxen called gaur and banteng have been gestated in cattle, but until now they have died soon after birth. Similar failures have been achieved in wild moufflon gestated in sheep; bongo antelope in eland antelope; Indian desert cat and African wild cat in domestic cats; and Grant's zebra in domestic horses. The failure of these zoo experiments suggests that a surrogate human mother could not carry a chimpanzee foetus to term. But they do at least prove that in every case, the baby comes out looking like its biological parent, not like its gestational parent. That, indeed, is the point of the experiment: to save rare species by mass-producing them in domestic animals' wombs.[33]

It is such an obvious outcome that the experiment seems pointless. We all know that a donkey embryo in a horse womb would develop into a donkey, not a horse. (Donkeys and horses are slightly more similar, genetically, than people and chimps. Like the two ape species, they also differ from each other in that horses have one more pair of chromosomes. This mismatch in chromosome number accounts for the sterility of mules and implies that a man mated to a female chimp just might produce a viable baby who would grow into a sterile ape-person with considerable hybrid vigour. Rumours of Chinese experiments in the 1950s notwithstanding, nobody seems to have tried this simple, but unethical experiment.)

So the conundrum only deepens. The genes, not the womb, determine our species. Yet despite having roughly the same set of genes, human beings and chimpanzees look different. How do you get two different species from one set of genes? How can we have a brain that is three times the size of a chimp's, and is capable of learning to speak, and yet not have an extra set of genes for making it?

THROWING SWITCHES

I cannot resist a literary analogy. The opening sentence of Charles Dickens's novel *David Copperfield* reads: 'Whether I shall turn out to be the hero of my own life, or whether that station will be held by anybody else, these pages must show.' The opening sentence of J.D. Salinger's novel *The Catcher in the Rye* reads: 'If you really want to hear about it, the first thing you'll probably want to know is where I was born, and what my lousy childhood was like, and how my parents were occupied and all before they had me, and all that David Copperfield kind of crap, but I don't feel like going into it.' In the pages that follow, to a close approximation, Dickens and Salinger use the same few thousand words. There are words that Salinger uses but not Dickens, like elevator or crap. There are words that Dickens uses but not Salinger, like caul and pettish. But they will be few compared with the words they share. Probably there is at least 90 per cent lexical concordance between the two books. Yet they are very different books. The difference lies not in the use of a different set of words, but in the same set of words used in a different pattern and order. Likewise, the source of the difference between a chimpanzee and a human being lies not in the different genes, but in the same set of 30,000 genes used in a different order and pattern.

I say this with confidence for one main reason. The most stunning surprise to greet scientists when they first lifted the lid on animal genomes was the discovery of the same sets of genes in wildly different animals. In the early 1980s, fly geneticists were thrilled to discover a small group of genes they called the hox genes that seemed to set out the body plan of the fly during its early development – roughly telling it where to put the head, the legs, the wings and so on. But they were completely unprepared for what came next. Their mouse-studying colleagues found recognisably the same hox genes, in the same order, doing the same job. The same gene tells a mouse embryo where (but not how) to grow ribs as tells a fly embryo where to grow wings: you can even swap them between species. Nothing had prepared biologists for this shock. It meant, in effect, that the basic body plan of all animals had been worked out in the genome of a long-extinct ancestor that lived more than 600 million years before and preserved ever since in its descendants (and that includes you).

Hox genes are the recipes for proteins called 'transcription factors', which means that their job is to 'switch on' other genes. A transcription factor works by attaching itself to a region of DNA called a promoter.[34] In creatures such as flies and people (as opposed to bacteria, say), promoters consist of about five separate stretches of DNA code, usually upstream of the gene itself, sometimes downstream. Each of those sequences attracts a different transcription factor, which in turn initiates (or blocks) the transcription of the gene. Most genes will not be activated until several of their promoters have caught transcription factors. Each transcription factor is itself a product of another gene somewhere else in the genome. The function of many genes is therefore to help switch other genes on or off. And the susceptibility of a gene to being switched on or off depends on the sensitivity of its promoters. If its promoters have shifted, or changed sequence so that the transcription factors find them more easily, the gene may be more active. Or if the change has made the promoters attract blocking transcription factors rather than enhancing ones, the gene may be less active.

Small changes in the promoter can therefore have subtle effects on the expression of the gene. Perhaps promoters are more like thermostats than switches. It is here in the promoters that scientists expect to find most evolutionary change in animals and plants – in sharp contrast to bacteria. For example, mice have short necks and long bodies; chickens have long necks and short bodies. If you count the vertebrae in the neck and thorax of a chicken and a mouse, you will find that the mouse has 7 neck and 13 thoracic vertebrae; the chicken has 14 and 7 respectively. The source of this difference lies in one of the promoters attached to one of the hox genes, Hoxc8, a gene found in both mice and chickens whose job is to switch on other genes that lay down details of development. The promoter is a 200-letter paragraph of DNA and it has just a handful of letters different in the two species. Indeed, changes in as few as two of these letters may be enough to make all the difference. The effect is to delay the expression of the Hoxc8 gene slightly in the development of the chicken embryo. Since development of the vertebral column starts at the head, this means the chicken goes on making neck vertebrae longer than the mouse.[35] In the python, Hoxc8 is expressed right from the head and goes on being expressed for most of the body. So pythons consist of one long thorax – they have ribs all down the body.[36]

The beauty of the system is that the same gene can be reused in different places and at different times simply by putting a set of different promoters beside it. The 'eve' gene in fruit flies, for example, whose job is to switch on other genes during development, is switched on at least ten separate times during the fly's life, and it has eight separate promoters attached to it, three upstream of the gene and five downstream. Each of these promoters requires 10–15 proteins to attach to it to switch on expression of the eve gene. The promoters cover thousands of letters of DNA text. In different tissues, different promoters are used to switch on the gene. This, incidentally, seems to be one reason for the humiliating fact that plants usually have more genes than animals. Instead of reusing the same gene by adding a new promoter to it, a plant reuses a gene by duplicating the whole gene and changing the promoter in the duplicated version. The 30,000 human genes are probably used in at least twice as many contexts during development thanks to batteries of promoters.[37]

To make grand changes in the body plan of animals, there is no need to invent new genes, just as there is no need to invent new words to write an original novel (unless your name is Joyce). All you need to do is switch the same ones on and off in different patterns. Suddenly, here is a mechanism for creating large and small evolutionary changes from small genetic differences. Merely by adjusting the sequence of a promoter, or adding a new one, you could alter the expression of a gene. And if that gene is itself the code for a transcription factor, then its expression will alter the expression of other genes. Just a tiny change in one promoter will produce a cascade of differences for the organism. These changes might be sufficient to create a wholly new species without changing the genes themselves at all.[38]

In one sense, this is a bit depressing. It means that until scientists know how to find gene promoters in the vast text of the genome, they will not learn how the recipe of a chimpanzee differs from that of a person. The genes themselves will tell them little, and the source of human uniqueness will remain as mysterious as ever. But in another sense it is also uplifting, reminding us, more forcefully than ever, of a simple truth that is all too often forgotten, that bodies are not made, they grow. The genome is not a blueprint for constructing a body; it is a recipe for baking a body. The chicken embryo is marinaded for a shorter time in the Hoxc8 sauce than the mouse embryo. This is a metaphor I shall return to frequently in the book, for it is one of the best ways of explaining why nature and nurture are not opposed to each other, but work together.

As the hox story illustrates, DNA promoters express themselves in the fourth dimension: their timing is all. A chimp has a different head from a human being not because it has a different blueprint for the head, but because it grows the jaws for longer and the cranium for less long than does the human being. The difference is all timing.

The process of domestication, by which the wolf was turned into the dog, illustrates the role of promoters. In the 1960s, a geneticist named Dmitri Belyaev was running a huge fur farm near Novosibirsk in Siberia. He decided to try to breed tamer foxes, because however well they had been handled and however many generations they had been kept in captivity, foxes were nervous and shy creatures in the fur farm (with good reason, presumably). So Belyaev started by selecting as breeding stock the animals that allowed him closest before fleeing. After 25 generations he did indeed have much tamer foxes, which, far from fleeing, would approach him spontaneously. The new breed of foxes not only behaved like dogs, they looked like dogs: their coats were piebald, like collies, their tails turned up at the end, the females came on heat twice a year, their ears were floppy, their snouts shorter and their brains smaller than in wild foxes. The surprise was that merely by selecting tameness, Belyaev had accidentally achieved all the same features that the original domesticator of the wolf had got – and that was probably some race of the wolf itself, which had bred into itself the ability not to run away too readily from ancient human rubbish dumps when disturbed. The implication is that some promoter change had occurred which affected not one, but many genes. Indeed, it is fairly obvious that what happened in both cases was that the timing of development had been altered so that the adult animals retained many of the features and habits of pups: the floppy ears, the short snout, the smaller skull and the playful behaviour.[39]

What seems to happen in these cases is that young animals do not yet show either fear or aggression, these developing last during the forward growth of the limbic system at the base of the brain. So the most likely way for evolution to produce a friendly or tame animal is to stop brain development prematurely. The effect is a smaller brain and especially a smaller 'area 13', a late-developing part of the limbic system that seems to have the job of disinhibiting adult emotional reactions such as fear and aggression. Intriguingly, such a taming process seems to have happened naturally in bonobos since their separation from the chimpanzee more than two million years ago. For its size the bonobo not only has a small head, but also reduced aggression and several juvenile features retained into adulthood including a white anal tail tuft, high-pitched calls and unusual female genitals. Bonobos have unusually small area 13s.[40]

So do human beings. Surprisingly, the fossil record suggests that there has been a rather steep decline in human brain size during the past 15,000 years, partly but not wholly reflecting a shrinking body size that seems to have accompanied the arrival of dense and 'civilised' human settlement. This followed several million years of more or less steady increases in brain size. In the Mesolithic (around 50,000 years ago) human brains averaged 1,468 cc (in females) and 1,567 cc (in males). Today the numbers have fallen to 1,210 cc and 1,248 cc, and even allowing for some reduction in body weight, this seems to be a steep decline. Perhaps there has been some recent taming of the species. If so, how? Richard Wrangham believes that once human beings became sedentary, living in permanent settlements, they could no longer tolerate anti-social behaviour and they began to banish, imprison or execute especially difficult individuals. In the past in highland New Guinea, more than one in ten of all adult deaths were by the execution of 'witches' (mostly men). This might have meant killing the more aggressive and impulsive – hence more developmentally mature and bigger-brained – people.[41]

Such self-taming, however, seems to be a recent phenomenon in our species and is not able to explain the selective pressures that led to the divergence of human beings from chimp-like ancestors more than five million years ago. But it does support the idea of evolution happening through the adjustment of gene promoters rather than genes themselves: hence the alteration of several irrelevant features caught in the slipstream of a reduction in impulsive aggression.[42] Meanwhile, it is suddenly looking possible to understand how the human brain achieved its enlarged size in the first place, thanks to a newly discovered gene on chromosome 1. Following the completion of a dam in Mirpur, in Pakistani-controlled Kashmir in 1967, a large number of local people, displaced from their homes, migrated to Bradford in England. They included some who had married cousins, and among the offspring of these cousin marriages were a few people born with abnormally small, though otherwise normal brains – so-called microcephalics. The family pedigrees allowed scientists to pin down the cause as four different mutations in different families, but all affecting the same gene: the ASPM gene on chromosome 1.

On further investigation, a team of scientists led by Geoffrey Woods in Leeds discovered something rather extraordinary about the gene. It is a large gene, 10,434 letters long and split into 28 paragraphs (called exons). The 16th to 25th paragraphs contain a characteristic motif repeated over and over again. The phrase, usually 75 letters long, begins with the code for the amino acids isoleucine and glutamine, the significance of which I will reveal in a moment. In the human version of the gene, there are 74 such motifs, in the mouse 61, in the fruit fly 24 and in the nematode worm just 2 repetitions. Remarkably, these numbers seem to be in proportion to the number of neurons in the adult brain of the animal.[43] Even more remarkably, the standard abbreviation for isoleucine is 'I' and the abbreviation for glutamine is 'Q'. Therefore, the number of IQ repeats may determine the relative IQ of the species, which, according to Woods, 'is a proof of God's existence since only someone with a sense of humour could have arranged for the correlation'.[44]

ASPM seems to work by regulating the number of times neuronal stem cells divide inside the vesicles of the young brain about two weeks after conception. This in turn decides how many neurons the adult brain will have. To have stumbled on a gene with the power to decide brain size in such a simple manner seems almost too good to be true, and complications will undoubtedly crowd in upon this simple story as more comes to be known. But the ASPM gene vindicates that young man who was so startled by the Fuegians: evolution is a difference of degree, not kind.

The startling new truth that has emerged from the human genome – that animals evolve by adjusting the thermostats on the fronts of genes, enabling them to grow different parts of their bodies for longer – has profound implications for the nature–nurture debate. Imagine the possibilities in a system of this kind. You can turn up the expression of one gene, the product of which turns up the expression of another, which suppresses the expression of a third, and so on. And right in the middle of this little network, you can throw in the effects of experience. Something external – education, food, a fight, or requited love, say – can influence one of the thermostats. Suddenly nurture can start to express itself through nature.

CHAPTER TWO

A plethora of instincts

When, as by a miracle, the lovely butterfly bursts from the chrysalis full-winged and perfect . . . it has, for the most part, nothing to learn, because its little life flows from its organization like melody from a music box.

Douglas Alexander Spalding, 1873[1]

Like Charles Darwin, William James was a man of independent means. He inherited a private income from his father Henry, whose father William had amassed $10,000 a year from the Erie Canal. The one-legged Henry used his self-sufficiency to become an intellectual, and spent much of his life shuttling between New York, Geneva, London and Paris with his children in tow. He was articulate, religious and self-assured. His two youngest sons went off to fight in the Civil War, then failed in business and turned to drink or depression. His two eldest sons, William and Henry, were trained almost from birth to be intellectuals. The result was (in Rebecca West's phrase) that 'one of them grew up to write fiction as though it were philosophy and the other to write philosophy as though it were fiction'.[2]

Both brothers were influenced by Darwin. Henry's novel *The Portrait of a Lady* was written in thrall to Darwin's idea of female choice as a force in evolution.[3] William's *Principles of Psychology*, much of which was first published as a series of articles in the 1880s, contained a manifesto for nativism – the idea that the mind cannot learn unless it has the rudiments of innate knowledge – going against the prevailing fashion for empiricism, the theory that behaviour is shaped by experience. William James believed that human beings were equipped with innate tendencies that were not derived from experience but from the Darwinian process of natural selection. 'He denies experience!' wrote James, quoting an imaginary reader. 'Denies science; believes the mind created by miracle; is a regular old partisan of innate ideas! That is enough! We'll listen to such antediluvian twaddle no more.'

William James asserted that human beings have more instincts than other animals, not fewer. 'Man possesses all the impulses that [lower creatures] have, and a great many more besides . . . It will be observed that no other mammal, not even the monkey, shows so large an array'. He argued that it was false to oppose instinct to reason:

> Reason, per se, can inhibit no impulses; the only thing that can neutralize an impulse is an impulse the other way. Reason may, however, make an inference which will excite the imagination so as to set loose the impulse the other way; and thus, though the animal richest in reason might also be the animal richest in instinctive impulses, too, he would never seem the fatal automaton which a merely instinctive animal would be.[4]

This is an extraordinary passage, not least because its impact on early twenty-first-century thought can be said to be almost nil. Very few people, on the side of either nature or nurture, took up such an extreme nativist position in the century to come, and almost everybody assumed for the following hundred years that reason was indeed the opposite of instinct. Yet James was no fringe lunatic. His work has influenced generations of scholars on consciousness, sensation, space, time, memory, will, emotion, thought, knowledge, reality, self, morality and religion – to name just the chapter headings of a modern book about his work. So why does this same book of 628 pages not even have the words 'instinct', 'impulse' or 'innate' in its index?[5] Why, for more than a century, has it been considered little short of indecent even to use the word 'instinct' in the context of human behaviour?

James's ideas were indeed immensely influential at first. His follower, William McDougall, founded a whole school of instinctivists, who became adept at spotting new human instincts for every circumstance. Too adept: speculation outstripped experiment and before long a counter-reformation was inevitable. In the 1920s the very empiricist ideas attacked by James, embodied in the notion of the blank slate, swept back to power not just in psychology (with John B. Watson and B.F. Skinner), but in anthropology (Franz Boas), psychiatry (Freud) and sociology (Durkheim). Nativism was almost totally eclipsed until 1958, when Noam Chomsky once again pinned its charter to the door of science. In a famous review of a book on language by Skinner, Chomsky argued that it was impossible for a child to learn the rules of language from examples: the child must have innate rules to which the vocabulary of the language was fitted. Even then, the blank slate dominated human sciences for many years. It was not until a century after his book was published that William James's idea of uniquely human instincts was at last taken seriously again in a new manifesto of nativism, written by John Tooby and Leda Cosmides (see chapter 9).

More of that later. First, a digression on teleology. It was Darwin's genius to turn the old theological argument from design on its head. Until then, the obvious fact that parts of organisms appear to be engineered for a purpose – the heart for pumping, the stomach for digesting, the hand for grasping – seemed logically to imply a designer, just as a steam engine implied the existence of an engineer. Darwin saw how the entirely backward-looking process of natural selection could none the less produce purposeful design – what Richard Dawkins called the blind watchmaker.[6] Though in theory it makes teleological nonsense to talk of a stomach having its own purpose, since the stomach has no mind, in practice it makes perfect sense so long as you engage the grammatical equivalent of four-wheel drive, the passive voice: stomachs have been selected to appear as if equipped with purposeful design. Since I have an aversion to the passive voice, I intend to avoid this problem throughout this book by pretending that there is indeed a teleological engineer thinking ahead and planning purposefully. The philosopher Daniel Dennett calls such an artefact a 'skyhook',[7] since it is the rough equivalent of a civil engineer hanging his scaffolding from the sky, but for the sake of simplicity I shall call my skyhook the Genome Organising Device, or GOD for short. This may keep religious readers happy, and allows me to use the active voice. So the question is: how does the GOD build a brain that can express an instinct?

Back to William James. To support his assertion that human beings have more instincts than other animals, James systematically enumerated the human instincts. He began with the actions of babies: sucking, clasping, crying, sitting up, standing, walking and climbing were all, he suggested, expressions of impulse, not imitations or associations. So, as the child grew, were emulation, anger and sympathy. So was a fear of strangers, loud noises, heights, the dark, reptiles. ('The ordinary cock-sure evolutionist ought to have no difficulty in explaining these terrors,' wrote James, neatly anticipating the argument of what is now called evolutionary psychology, 'as relapses into the consciousness of the cave-men, a consciousness usually overlaid in us by experiences of more recent date.') He moved on to acquisitiveness, noting the tendency of boys to collect things. He noticed the very different play preferences of boys and girls. Parental love, he suggested, was at least initially stronger in women than in men. He tripped quickly through sociability, shyness, secretiveness, cleanliness, modesty and shame. 'Jealousy is unquestionably instinctive,' he remarked.

The strongest of the instincts, he believed, was love. 'Of all propensities, the sexual impulses bear on their face the most obvious signs of being instinctive, in the sense of blind, automatic and untaught.'[8] But, he insisted, just because sexual attraction was instinctive did not mean it was irresistible. Other instincts, like shyness, prevent us acting upon every sexual attraction.

So let me take James at his word, provisionally at least, and examine the idea of the love instinct in a little more depth. If he is right, there must be some heritable factor, which gives rise to a physical or chemical change in our brains when we fall in love, that change causing, rather than caused by, the emotion of falling in love. Such as this, from the scientist Tom Insel:

> A working hypothesis is that oxytocin released during mating activates those limbic sites rich in oxytocin receptors to confer some lasting and selective reinforcement value on the mate.[9]

Or, to put it more poetically, you fall in love.

What is this oxytocin and why does Insel make such an extravagant claim for it? The story starts in an almost ridiculously unromantic process: urination. Some 400 million years ago, when the ancestors of our species first left the water, they were equipped with a tidy little hormone called vasotocin, a miniature protein made out of a chain of just nine amino acids formed into a ring. Its job was to regulate salt and water balance in the body, and it performed this job by rushing about switching on cells in the kidney or other organs. Fish still use two different versions of vasotocin for this purpose today, and so do frogs. In the descendants of reptiles – and that includes human beings – there are two slightly different copies of the relevant gene lying next to each other, facing different ways (in human beings on chromosome 20). The result today is that all mammals have two such hormones, called vasopressin and oxytocin, that differ at two of the links in the chain.

They still do their old job. Vasopressin tells the kidney to conserve water; oxytocin tells it to excrete salt. But, like vasotocin in modern fish, they also have a role in the regulation of reproductive physiology. Oxytocin stimulates the contraction of muscles in the womb during birth; it also causes milk to be expelled from the ducts in the breast. The GOD is an economiser: having invented a switch for one purpose, he readapts it for other purposes, by expressing the oxytocin receptor in a different organ. But a much greater surprise came in the early 1980s, when scientists suddenly realised that vasopressin and oxytocin had a job to do inside the brain as well as being secreted from the pituitary gland into the bloodstream.

So they tried injecting oxytocin and vasopressin into the brains of rats to see what effect they had. Bizarrely, a male rat injected with intracerebral oxytocin immediately begins yawning and simultaneously gets an erection.[10] So long as the dose is low, the rat also becomes more highly sexed: it ejaculates sooner and more frequently. In female rats, intracerebral oxytocin induces the animal to adopt a mating posture. In human beings, meanwhile, masturbation increases oxytocin levels in both sexes. All in all, oxytocin and vasopressin in the brain seem to be connected to mating behaviour.

Now all of this sounds rather unromantic: urine, masturbation, breastfeeding – hardly the essence of love. Be patient. In the late 1980s, Tom Insel was working on the effect of oxytocin on maternal behaviour in rats. Brain oxytocin seemed to help the mother rat form a bond with its young and Insel identified the parts of the rat brain that were sensitive to the hormone. He switched his attention to the pair bond, wondering if there were parallels between a female's bond to her young and to her mate. At this point he met Sue Carter, who had begun to study prairie voles in the laboratory. She told him how the prairie vole is a rarity among mice for its faithful marriages. Prairie voles live in couples and both father and mother care for the young for many weeks. Montane voles, on the other hand, are more typical of mammals: the female mates with a passing polygamist, separates quickly from him, bears young alone and abandons them after a few weeks to fend for themselves. Even in the laboratory, this difference is clear: mated prairie voles stare into each other's eyes and bathe the babies; mated montane voles treat their spouses like strangers.

Insel examined the brains of the two species. He found no difference in the expression of the two hormones themselves, but a big difference in the distribution of molecular receptors for them – the molecules that fire up neurons in response to the hormones. The monogamous prairie voles had far more oxytocin receptors in several parts of the brain than the polygamous montane voles. Moreover, by injecting oxytocin or vasopressin into the brains of prairie voles, Insel and his colleagues could elicit all the characteristic symptoms of monogamy, such as a strong preference for one partner and aggression towards other voles. The same injections had little effect on montane voles, and the injection of chemicals that block the oxytocin receptors prevented the monogamous behaviour. The conclusion was clear: prairie voles are monogamous because they respond more to oxytocin and vasopressin.[11]

In a virtuoso display of scientific ingenuity, Insel's team has gone on to dissect this effect in convincing detail. They knock the oxytocin gene out of a mouse before birth. This leads to social amnesia: the mice can remember things, but they have no memory of mice they have already met and will not recognise them. Lacking oxytocin in its brain, a mouse cannot recognise a mouse it has just met ten minutes before – unless that mouse was 'badged' with a non-social cue such as a distinctive lemon- or almond-scented smell (Insel compares this to an absent-minded professor at a conference who recognises friends by their name tags, not their faces).[12] Then by injecting the hormone into just one part of the animal's brain in later life – the medial amygdala – the scientists can restore social memory to the mouse completely.

In another experiment, using a specially adapted virus, they turn up the expression of the vasopressin receptor gene in the ventral pallidum, a part of a vole's brain important for reward. Pause here to roll that idea around your mind a few times to appreciate just what science can do these days: they use viruses to turn up the volumes of genes in one part of the brain of a rodent. Even ten years ago such an experiment was unimaginable. The result of turning up the gene's expression is to 'facilitate partner preference formation', which is geekspeak for 'make them fall in love'. They conclude that for a male vole to pair-bond, it must have both vasopressin and vasopressin receptors in its ventral pallidum. Since mating causes a release of oxytocin and vasopressin, the prairie vole will pair-bond with whatever animal it has just mated; the oxytocin helps in memory, the vasopressin in reward. The montane vole, by contrast, will not react in the same way, because it lacks receptors in that area. Female montane voles express these receptors only after giving birth, so they can be nice to their babies, briefly.

So far I have talked of oxytocin and vasopressin as if they were the same thing, and they are so similar that they probably stimulate each other's receptors somewhat. But it appears that to the extent that they do differ, oxytocin makes female voles choose a partner; vasopressin makes males choose a partner. The male prairie vole becomes aggressive towards all voles except its mate when vasopressin is injected into his brain. Attacking other voles is a (rather male) way of expressing his love.[13]

All this is astonishing enough, but perhaps the most exciting result to emerge from Insel's lab concerns the genes for the receptors. Remember that the difference between the prairie vole and the montane vole lies not in the expression of the hormone, but in the pattern of expression of the hormone's receptors. These receptors are themselves the products of genes. The receptor genes are essentially identical in the two species, but the promoter regions, upstream of the genes, are very different. Now recall the lesson of chapter 1: that the difference between closely related species lies not in the text of genes themselves, but in their promoters. In the prairie vole, there is an extra chunk of DNA text, on average about 460 letters long, in the middle of the promoter. So Insel's lab made a transgenic mouse with this expanded promoter and it grew up with a brain like a prairie vole, expressing vasopressin receptors in all the same places, though it did not form a pair bond.[14] Steven Phelps then went out and caught 43 wild prairie voles in Indiana and sequenced their promoters: some had longer insertions than others. They varied from 350 to 550 letters in length. Are the long ones more faithful husbands than the short ones? Not yet known.[15]

The conclusion to which Insel's work is leading is devastating in its simplicity. The ability of a rodent to form a long-term attachment to its sexual partner may depend on the length of a piece of DNA text in the promoter switch at the front of a certain receptor gene. That in turn decides precisely which parts of the brain will express the gene. Of course, like all good science, this discovery raises more questions than it settles. Why should feeding oxytocin receptors in that part of the brain make the mouse feel well-disposed to its partner? It is possible that the receptors induce a state a bit like addiction, and in this respect it is noticeable that they seem to link with the D2 dopamine receptors, which are closely involved in various kinds of drug addiction.[16] On the other hand, without oxytocin, mice cannot form social memories, so perhaps they simply keep forgetting what their spouse looks like.

Mice are not men. You know by now that I am about to start extrapolating anthropomorphically from pair-bonding in voles to love in people, and you probably do not like my drift. It sounds reductionist and simplistic. Romantic love, you say, is a cultural phenomenon, overlaid with centuries of tradition and teaching. It was invented at the court of Eleanor of Aquitaine, or some such place, by a bunch of oversexed poets called troubadours; before that there was just sex.

Even though in 1992 William Jankowiak surveyed 168 different ethnographic cultures and found none that did not recognise romantic love, you may be right.[17] I certainly cannot prove to you yet that people fall in love when their oxytocin and vasopressin receptors get tingled in the right places in their brains. Yet. And there are cautionary hints about the dangers of extrapolating from one species to another: sheep seem to need oxytocin to form maternal attachment to their young; mice apparently do not.[18] Human brains are undoubtedly more complicated than mouse brains.

But I can draw your attention to some curious coincidences. A mouse shares much of its genetic code with a human being. Oxytocin and vasopressin are identical in the two species and are produced in the equivalent parts of the brain. Sex causes them to be produced in the brain in both human beings and rodents. Receptors for the two hormones are virtually identical and are expressed in equivalent parts of the brain. Like those of the prairie vole, the human receptor genes (on chromosome 3) have a – smaller – insertion in their promoter regions. Like the prairie voles of Indiana, the lengths of those promoter insertions vary from individual to individual: in the first 150 people examined, Insel found 17 different promoter lengths. And when a person who says she (or he) is in love contemplates a picture of her loved one while sitting in a brain scanner, certain parts of her brain light up that do not light up when she looks at a picture of a mere acquaintance. Those brain parts overlap with the ones stimulated by cocaine.[19] All this could be a complete coincidence, and human love may be entirely different from rodent pair-bonding, but given how conservative the GOD is and how much continuity there is between human beings and other animals, you would be unwise to bet on it.[20]

Shakespeare was ahead of us, as usual. In *A Midsummer Night's Dream*, Oberon tells Puck how Cupid's arrow fell upon a white flower (the pansy), turning it purple, and that now the juice of this flower

> . . . on sleeping eyelids laid
> Will make or man or woman madly dote
> Upon the next live creature that it sees.

Puck duly fetches a pansy and Oberon wreaks havoc with the lives of those sleeping in the forest, causing Lysander to fall in love with Helena, whom he has previously scorned, and causing Titania to fall in love with Bottom the weaver wearing the head of an ass.

Who would now wager against me that I could not soon do something like this to a modern Titania? Admittedly, a drop on the eyelids would not suffice. I would have to give her a general anaesthetic while I cannulated her medial amygdala and injected oxytocin into it. I doubt even then that I could make her love a donkey. But I might stand a fair chance of making her feel attracted to the first man she sees upon waking. Would you bet against me? (I hasten to add that ethics committees will – should – prevent anybody taking up my challenge.)

I am assuming that, unlike most mammals, human beings are basically monogamous, like prairie voles, and not promiscuous, like montane voles. I base this assumption on the testicle-size argument enunciated in chapter 1; on the ample evidence from ethnography that, though most human societies allow polygamy, most human societies are still dominated by monogamous relationships; and on the fact that human beings usually practise some paternal care – a characteristic feature of the few mammal species that live as social monogamists.[21] Furthermore, as we have liberated human life from economic and cultural straitjackets, such as arranged marriage, we have found monogamy growing more dominant, not less. In 1998 the most powerful man in the world, far from treating himself to a gigantic harem, got into trouble for having an affair with one intern. The evidence for long-term, exclusive (but sometimes cheated-on) pair bonds as the commonest pattern in human relationships is all around you.

Chimpanzees are different. Long-term pair bonds are unknown, and I predict that they have fewer oxytocin receptors in the relevant parts of their brains than human beings, probably as a result of having shorter gene promoters. The oxytocin story lends at least tentative support to William James's notion that love is an instinct, evolved by natural selection, and is part of our mammal heritage, just like four limbs and ten fingers. Blindly, automatically and untaught, we bond with whoever is standing nearest when the oxytocin receptors in the medial amygdala get tingled. One sure way to tingle them is to have sex, although presumably chaste attraction can also do the trick. Is this why breaking up is hard to do?

Having oxytocin receptors does not make it inevitable that somebody will fall in love during his life, nor predictable when it will happen, or with whom. As Niko Tinbergen, the great Dutch ethologist, demonstrated in his studies of instincts, the expression of a fixed, innate instinct must often be triggered by an external stimulus. One of Tinbergen's favourite species was the stickleback, a tiny fish. Male sticklebacks go red on the belly in the breeding season, when they defend small territories in which they build nests, which attract females. Tinbergen made little models of fish and caused them to 'invade' the territory of a male fish. A model of a female elicited the courtship dance of the male, even if the model was astonishingly crude; so long as it had a 'pregnant' belly, it excited the male. But if the model had a red belly, it would trigger an attack. It could be just an oval blob with a crudely drawn eye but no fins or tail: still it was attacked just as vigorously as if it were a real male rival – so long as it was red. One of the legends of Leiden, where Tinbergen first worked, is that he noticed his sticklebacks would threaten the red post-office vans that drove past the window.

Tinbergen went on to demonstrate the power of these 'innate

releasing mechanisms' to provoke an instinct in other species, notably the herring gull. Herring gulls have yellow beaks with a bright red spot near the tip. The chicks peck at this spot when begging for food. By presenting newborn chicks with a series of models, Tinbergen demonstrated that the spot was a powerful releaser for the begging action, and the redder it was the better. The colour of the beak or the head of the bird mattered not at all. So long as it had a contrasting spot near the tip of the bill, preferably in red, it would elicit pecking. In modern jargon, scientists would say that the chick's instinct, and the adult's beak spot had 'co-evolved'. An instinct is designed to be triggered by an external object or event. Nature plus nurture.[22]

The significance of Tinbergen's experiments was to reveal just how complex instincts could be, and yet how simply triggered. The digger wasp he studied would dig a burrow, go and catch a caterpillar, paralyse it with a sting, bring it back to the burrow and deposit it with an egg on top, so that the baby wasp could feed on the caterpillar while growing. All of this complex behaviour, including the ability to navigate back to the burrow, was achieved with almost no learning, let alone parental teaching. A digger wasp never meets its parents. A cuckoo migrates to Africa and back, sings its song and mates with one of its own species without as a chick ever seeing either a parent or a sibling.

The notion that animal behaviour is in the genes once troubled biologists as much as it now troubles social scientists. Max Delbruck, pioneering molecular biologist, refused to believe that his colleague at Caltech Seymour Benzer had found a behavioural mutant fly. Behaviour, he insisted was too complex to reduce to single genes. Yet the idea of behaviour genes has long been accepted by the amateur breeders of domestic animals. The Chinese started breeding mice of different colours in the seventeenth century or earlier and they produced a mouse called the waltzing mouse, famous for its dance-like gait caused by an inherited defect in the inner ear. Mouse breeding then caught on in Japan in the nineteenth century and thence spread to Europe and America. Some time before the year 1900 a retired schoolteacher in Granby, Massachusetts, by the name of Abbie

Lathrop, took up the 'mouse fancying' hobby. Soon she was breeding different strains of mice herself in a small barn adjoining her property and selling them to pet shops. She was especially fond of what were by then known as Japanese waltzing mice, and she developed several new strains. She also noticed that some strains got cancer more often than others; picked up by Yale University, this hint became the basis of early studies of cancer.

But it was Lathrop's link to Harvard that uncovered the link between genes and behaviour. William Castle of Harvard bought some of her mice and started a mouse laboratory. Under Castle's student Clarence Little, the main mouse laboratory moved to Bar Harbor, Maine, where it still is – a giant factory of inbred mouse strains used in research. Very early on, the scientists began to realise that different strains of mice behaved in different ways, too. Benson Ginsburg, for instance, found out the hard way. He noticed that when he picked up a mouse of the 'guinea-pig' strain (named for its coat colour), he often got bitten. He was soon able to breed a new strain that had the coat colour but not the aggressive streak: proof enough that aggression was somewhere in the genes. His colleague Paul Scott also developed aggressive strains of mice, but bizarrely, Ginsburg's most aggressive strain was Scott's most pacific. The explanation was that Scott and Ginsburg had handled the mice differently as babies. For some strains, handling did not matter. But for one strain in particular, C57-Black-6, early handling increased the aggressiveness of the mouse. Here was the first hint that a gene must interact with an environment if it is to have its effect. Or, as Ginsburg put it, the road from the 'encoded genotype' the mouse inherits to the 'effective genotype' it expresses passes through the process of social development.[23]

Ginsburg and Scott both later went on to work with dogs, Scott proving by crossing experiments between cocker spaniels and African basenjis that play-fighting in puppies is controlled by two genes that regulate the threshold for aggression.[24] But it did not need science to prove the inheritance of behaviour in dogs: that was old news to dog-breeders. The point of dogs is that they come in different behavioural types: retrievers, pointers, setters, shepherds, terriers, poodles, bulldogs, wolfhounds – their very names denote the fact that they have instincts bred into them. And those instincts are innate. A retriever cannot be trained to guard livestock and a guard dog cannot be trained to herd sheep. It's been tried. In the process of domestication, dogs have kept incomplete or exaggerated elements of wolf behaviour development. A wolf will stalk, chase, pounce, grab, kill, dissect, and carry food, and a wolf pup will practise each of these activities in turn as it grows up. Dogs are wolf pups frozen in the practising stage. Collies and pointers are stuck in the stalking stage; retrievers are stuck with carrying and pit bulls with biting: each is a frozen mixture of different wolf-pup themes. Is it in their genes? You bet: 'Breed-specific behaviours are irrefutable,' says dog chronicler Stephen Budiansky firmly.[25]

Or go ask the cattle-breeders. I have in front of me a catalogue of dairy bulls designed to entice me into ordering some semen by mail. In enormous detail it describes the quality and shape of the bull's udder and teats, its milk-producing ability, its milking speed and even its temperament. But surely, you point out, bulls don't have udders? On every page there is a picture of a cow, not a bull. What the catalogue is referring to is not the bull himself but his daughters. 'Zidane, the Italian No 1,' it boasts, 'improves frame traits and fixes on tremendous rumps with ideal slope. He is particularly impressive in his feet and leg composites with excellent set and terrific depth of heel. He leaves faultless udders, which are snugly attached with deep clefts.' The characters are all female, but the attribution is to the sire. Perhaps I would prefer to buy a straw of semen from Terminator, whose daughters have 'great teat placement', or Igniter, a bull that is a 'milking speed specialist' whose daughters 'display great dairy character'. I might wish to avoid Moet Flirt Freeman, because although his daughters have 'tremendous width across the chest' and give more milk than their mothers did, the small print admits they are also slightly 'below average' in temperament – which probably means that they tend to kick out when being milked. They are also slow milkers.[26]

The point is that cattle-breeders have no qualms about attributing behaviour to genes, just as they attribute anatomy to genes. Minute differences in the behaviour of cows they confidently ascribe to the semen that arrived through the mail. Human beings are not cows. Admitting instinct in cows does not prove that human beings are also ruled by instinct, of course. But it demolishes the assumption that because behaviour is complex or subtle, it cannot be instinctive. Such a comforting illusion is still rife within the social sciences; yet no zoologist who has studied animal behaviour could believe that complex behaviour cannot be innate.

MARTIANS AND VENUSIANS

Defining 'instinct' has baffled so many scientists that some refuse to use the word altogether. It need not be present from birth: some instincts only develop in adult animals (as wisdom teeth do). It need not be inflexible: digger wasps will alter their behaviour according to how many caterpillars they find already in the burrow they are provisioning. It need not be automatic: unless it meets a red-bellied fish, the stickleback male will not fight. And the boundaries between instinctive and learned behaviour are blurred.

But imprecision does not necessarily render a word useless. The boundaries of Europe are uncertain – how far east does it stretch? Are Turkey and Ukraine in it? – and there are many different meanings of the word 'European', but it is still a useful word. The word 'learn' covers a multitude of virtues, but it is still a useful word. Likewise, I believe that to call behaviour instinctive can still be useful. It implies that the behaviour is at least partially inherited, hard-wired and automatic, given the expected environment. A characteristic feature of an instinct is that it is universal. That is, if something is primarily instinctive, then it must be approximately the same in all people. Anthropology has always been torn between an interest in human similarities and human differences, with the advocates of nature emphasising the former and the advocates of nurture stressing the latter. The fact that people smile, frown, grimace and laugh in much the same way all over the world struck Darwin and would later strike the ethologists Irenaeus Eibl-Eibesfeldt and Paul Ekman as astonishing. Even among those inhabitants of New Guinea and the Amazon till then uncontacted by 'civilisation', these emotional expressions have the same form and the same meaning.[27] At the same time, the astonishing variety of rituals and habits expressed by the human race testifies to its capacity for difference. As usual in science, each side of the argument pushed the other to extreme positions.

Perhaps it would satisfy both (or neither) to focus on the paradox of human differences that are universally similar all over the world. After all, similarity is the shadow of difference. The prime candidate is sex and gender difference. Nobody now denies that men and women are different not just in anatomy but also in behaviour. From best-selling books about them being from different planets to the increasing polarisation of films into those that appeal to men (action) or to women (relationships), it is surely no longer controversial to assert that – despite exceptions – there are consistent mental as well as physical differences between the sexes. As the comedian Dave Barry puts it, 'If a woman has to choose between catching a fly ball and saving an infant's life, she will choose to save the infant's life without even considering if there are men on base.' Are such differences nature, nurture, or both?

Of all the sex differences, the best studied are the ones to do with mating. In the 1930s, psychologists first started asking men and women what they sought in a mate, and they have been asking them ever since. The answer seems so obvious that only a laboratory nerd or a Martian would bother to ask the question. But sometimes the most obvious things are the ones that most need demonstrating.

They found many similarities. Both sexes wanted intelligent, dependable, cooperative, trustworthy and loyal partners. But they also found differences. Women rated good financial prospects in their partners twice as highly as men. Hardly surprising, since men were breadwinners in the 1930s. Come back in the 1980s and you would surely find such a patently cultural difference vanishing. No: in every survey conducted since then, right up to the present day, the same preference emerges just as strongly. To this day, American women rate financial prospects twice as highly as men do when seeking mates. In personal advertisements, women mention wealth as a desirable feature of a partner 11 times as often as men do. The psychology establishment dismissed this result: it merely reflected the importance of money in American culture, not a universal sex difference. So the psychologist David Buss went and asked foreigners, and got the same answer from Dutch and German men and women. Don't be absurd, he was told, Western Europeans are just like Americans. So Buss asked 10,047 people from 37 different cultures on six continents and five islands, ranging from Alaska to Zululand. In every culture, bar none, women rated financial prospects more highly than men. The difference was highest in Japan and lowest in Holland but it was always there.[28]

This was not the only difference he found. In all 37 cultures, women wanted men older than them. In nearly all cultures, social status, ambition and industriousness in a mate mattered more to women than to men. Men by contrast placed more emphasis on youth (in all cultures, men wanted younger women) and physical appearance (in all cultures, men wanted beautiful women more than women wanted beautiful men). In most cultures, men also placed slightly more emphasis on chastity and fidelity in their partners, while (of course) being much more likely to seek extramarital sex themselves.[29]

Well, what a surprise! Men like pretty, young, faithful women, while women like rich, ambitious, older men. A casual glance through films, novels or the newspaper could have revealed this to Buss, or any passing Martian. Yet the fact remains that many psychologists had firmly told Buss he would not be able to find such trends repeated outside the countries of the West, let alone all over the world. Buss proved something which was – at least to the social science establishment – very surprising.

Many social scientists argue that the reason women seek wealthy men is that men have most of the wealth. But now you know this is universal to the human race, you could easily turn it around. Men seek wealth because they know it attracts women – just as women pay more attention to appearing youthful because they know it attracts men.

This direction of causality was never less plausible than the other, and given the evidence of universality, it is now more plausible. Aristotle Onassis, who knew a bit about both money and beautiful women, reputedly once said: 'If women did not exist, all the money in the world would have no meaning.'[30]

By proving how universal so many sex differences in mating preferences are, Buss has thrown the burden of proof on to those who would see a cultural habit rather than an instinct. But the two explanations are not mutually exclusive. They are probably both true. Men seek wealth to attract women, therefore women seek wealth because men have it, therefore men seek wealth to attract women. And so on. If men have an instinct to seek the baubles that lead to success with women, then they are likely to learn that within their culture money is one such bauble. Nurture is reinforcing nature, not opposing it.

With the human species, as Dan Dennett observed, you can never be sure that what you see is instinct, because you might be looking at the result of a reasoned argument, a copied ritual or a learned lesson. But the same applies in reverse. When you see a man chasing a woman just because she is pretty, or a girl playing with a doll while her brother plays with a sword, you can never be sure that what you are seeing is just cultural, because it might have an element of instinct. Polarising the issue is entirely mistaken. It is not a zero-sum game, where culture displaces instinct or vice versa. There might be all sorts of cultural aspects to a behaviour that is grounded in instinct. Culture will often reflect human nature rather than affect it.

MONEY OR DIAMOND?

Buss's study of global similarity in difference proves the universality of different approaches to mating behaviour, but says nothing about how they come about. Suppose he is right and the differences are evolved, adaptive and therefore at least partly innate. How do they develop and under what influences? Thanks to an extraordinary battle in the nature–nurture war, called Money vs. Diamond, there is now a glimmer of light to be cast upon this subject.

Money is John Money, a psychologist from New Zealand who reacted against his strict religious upbringing to become an outspoken 'missionary' of sexual liberation at Johns Hopkins University in Baltimore, eventually defending not just free love but even consenting paedophilia. Diamond is Mickey Diamond, a tall, soft-spoken, bearded son of Ukrainian Jewish immigrants to the Bronx who moved first to Kansas and then to Honolulu, where he studies the factors determining sexual behaviour in animals and people.

Money believes that sex roles are the products of early experience, not instinct. In 1955 he set out his theory of psychosexual neutrality based on the study of 131 human 'hermaphrodites' – people who had been born with ambiguous genitalia. At birth, said Money, human beings are psychosexually neutral. Only after experience, at about the age of two, do they develop 'gender identity'. 'Sexual behavior and orientation as male or female does not have an innate, instinctive basis,' he wrote. 'It becomes differentiated as masculine or feminine in the course of the various experiences of growing up.' Therefore, said Money, a human baby can be literally assigned to either sex, a belief that was used by doctors to justify surgery to change baby boys born with abnormal penises into girls. Such surgery became standard practice: males with unusually tiny penises were 'reassigned' as females.

In contrast, the Kansas group came to the conclusion that 'the biggest sex organ is between the ears, not between the legs' and began to challenge the orthodoxy that sex roles were environmentally determined. In 1965 Diamond argued the point in a paper critical of Money, charging that Money had presented no case histories to support his theory of psychosexual neutrality, that the evidence from hermaphrodites was irrelevant – if their genitalia were ambiguous, their brains might be, too – and that it was more plausible that human beings, like guinea pigs, experienced a prenatal fixation of mental sex identity.[31] In effect, he challenged Money to produce a psychosexually neutral, normal child, or one that had accepted sex reassignment.

Money brushed aside the criticism as he gathered the rewards of increasing fame. His paper had won a prize; that had led to a huge grant; and when his team began transsexual surgery, he became a celebrity profiled in newspapers and on television. But Diamond had hit a nerve, for the very next year Money took on a case of a normal boy who had lost his penis after a botched circumcision. The boy was a monozygotic twin, so the opportunity to demonstrate how he could be turned into a woman, while his twin developed as a man, was irresistible. On Money's advice the boy was surgically reassigned as a girl then raised by his parents as a girl and never told of her origin. In 1972 Money published a book describing the case as an unqualified success. It was hailed in the press as definitive proof that sex roles were the product of society, not biology; it influenced a generation of feminists at a critical time; it entered the psychology textbooks; and it influenced multitudes of doctors who now saw sex reassignment as a simple solution to a complicated problem.

Money seemed to have won the argument. Then in 1979 a BBC television team began investigating the case. They had heard rumours that the boy who became a girl was not the success that Money claimed. They managed to penetrate the anonymity of the case and even briefly meet the girl in question, though they did not divulge her identity on air. Called Brenda Reimer, she lived with her family in Winnipeg and was then 14. What they saw was an unhappy youth with masculine body language and a deep voice. The BBC crew interviewed Money, who reacted with fury at the invasion of the family's privacy. Diamond continued to press Money for details, but got nowhere. Money now dropped all reference to the case from his published work. The trail once more went cold. Then in 1991 in print Money blamed Diamond for inciting the BBC to invade the girl's privacy. Enraged at the accusation, Diamond began trying to contact psychiatrists who might have treated the case. In 1995 at last he met Brenda Reimer.

Except Brenda was now called David, and was a happily married man with adopted children. He had endured a confused and unhappy childhood, constantly rebelling against girlish things, though he knew nothing of having been born a boy. When at 14 he still insisted on living as a boy his parents at last told him of his past. He immediately demanded surgery to restore a penis and adopted the life of a teenage male. Diamond persuaded him to let him tell the story to the world under a pseudonym so that they could prevent people having to endure the same fate in the future. In 2000, the writer John Colapinto convinced him to drop his anonymity altogether for a book.[32]

Money has never apologised either to the world for misleading it about the success of the reassignment, or to David Reimer. Today Diamond wonders what would have happened if the little boy had been a gay or transsexual who might have wanted to live either in an effeminate way or as a female, or had not been willing to come out of his closet and tell his story.

David Reimer is not alone. Most boys reassigned as girls declare themselves boys at adolescence. And a recent study of people born with ambiguous genitalia found that those who escaped the surgeon's knife had fewer psychological problems than those who had been operated on in childhood. The large majority of those males that were switched to live as girls have reverted, on their own, to live as males.[33]

Gender roles are at least partly automatic, blind and untaught, to use William James's terms. Hormones within the womb trigger masculinisation, but those hormones originate within the body of the baby and are themselves triggered by a series of events that begin with the expression of a single gene on the Y chromosome. (There are plenty of species that allow the environment to determine gender. In crocodiles and turtles, for example, the sex of the animal is set by the temperature at which the egg is incubated. But there are genes involved in such a process, too. Temperature triggers the expression of sex-determining genes. The prime cause may be environmental, but the mechanism is genetic. Genes can be consequence as well as cause.)

FOLK PSYCHOLOGY

Boys like David Reimer want to be boys. They like toys, weapons, competition and action better than dolls, romance, relationships and families. They do not come into the world with all these preferences fully formed, of course, but they do come with some ineffable preference to identify with boyish things. This is what the child psychologist Sandra Scarr has called 'niche picking': the tendency to pick the nurture that suits your nature. The frustrations of David Reimer's youth were caused by his not being allowed to pick his niche.

In this sense, cause and effect are probably circular. People both like doing what they find they are good at and are good at what they like doing. But that implies that this sex difference is at least jump-started by instinct, by innate behavioural differences that pre-date experience. Like many parents who have had children of both sexes, I found the differences surprisingly strong and early. I also had no difficulty in believing that I and my wife were reacting to, rather than causing, such gender dissimilarities. We bought trucks for the boy and dolls for the girl not because we wanted them to be different, but because it was painfully obvious that one wanted trucks and the other dolls.

Exactly how early do these differences emerge? Svetlana Lutchmaya, a student of Simon Baron-Cohen's at Cambridge, filmed 29 girls and 41 boys at 12 months old and analysed how often the baby looked at its mother's face. As expected, the girls made far more eye contact than the boys. She then went back and measured the testosterone levels present in the womb during the first trimester of each baby's gestation. This was possible because in every case the mother had had amniocentesis and a sample of amniotic fluid had been stored. She found that the foetal testosterone level was generally higher for the boys than the girls, and that, among the boys, there was a significant correlation: the higher the testosterone level, the less eye contact made by the baby as a one-year-old.[34]

Baron-Cohen then asked another student, Jennifer Connellan, to go back even further, to the first day of life. She gave 102 24-hour-old babies two things to look at: her own face, or a physical-mechanical mobile of approximately the same size and shape as a face. The baby boys slightly preferred to look at the mobile; the baby girls slightly preferred the face.[35]

So the relative female preference for faces, which gradually turns into a preference for social relationships, seems to be there in some form from the start. This distinction between the social and physical world may be a crucial clue to how human brains work. The nineteenth-century psychologist Franz Brentano divided the universe rather starkly into two kinds of entities: those that have intentionality and those that do not. The former can move themselves spontaneously and can have goals and wants; the latter obey only physical laws. It is a distinction that fails at the edges – what about plants? – but as a rule of thumb it works rather well. Evolutionary psychologists have begun to suspect that human beings instinctively apply two different mental processes to understanding such objects: what Daniel Dennett has called folk psychology and folk physics. We assume that a footballer moved because he 'wanted to', but that a football moved only because it was kicked. Even babies express surprise when objects appear to disobey the laws of physics – if they move through each other, if large objects seem to go into smaller ones, or if they move without being touched.

You can see where I am heading, I suspect: on average, men are more interested in folk physics than women, who are more interested in folk psychology than men. Simon Baron-Cohen's research focus is autism, a difficulty with the social world that affects mainly boys. Together with Alan Leslie, Baron-Cohen pioneered the theory that autistic boys have trouble theorising about the minds of others, though he now prefers to use the term 'empathising'. There are many other features of severe autism, including difficulty with language, but in what is probably its 'purer' and less severe form, Asperger's syndrome, autism seems mainly to consist of a difficulty in empathising with other people's thoughts. Since boys are less good at empathising than girls anyway, then perhaps autism is just an extreme version of the male brain. Hence Baron-Cohen's interest in the inverse correlation between prenatal testosterone and eye contact: the masculinisation of the brain by testosterone may go 'too far' in autistics.[36]

Intriguingly, Asperger's children are often better than normal at folk physics. Not only are they frequently fascinated by mechanical things, from light switches to aeroplanes, but they generally take an engineering approach to the world, trying to understand the rules by which things – and people – operate. They frequently become precociously expert in factual knowledge and mathematics. They are also more than twice as likely to have fathers and grandfathers who worked in engineering. On a standard test of autistic tendencies, scientists generally score higher than non-scientists and physicists and engineers score higher than biologists. Baron-Cohen says of one brilliant mathematician, a winner of the Fields medal, who has Asperger's: 'Empathy passes him by.'[37]

To demonstrate how a difficulty with folk psychology can coexist happily with expertise at folk physics, psychologists designed two remarkably similar tests called the false-belief test and the false-photo test. In the false-belief test, the child sees the experimenter move a concealed object from one receptacle to another while a third person is not watching. The child then has to say where the third person will look for the object. To get the right answer, he has to understand that the third person holds a false belief. All children pass this test for the first time around the age of four (boys later than girls), but autistics are especially late developers.

In the false-photo test, by contrast, the child takes a Polaroid photograph of a scene, then, while the picture is developing, sees the experimenter move one of the objects in the scene. The child is asked which position the object will occupy in the photograph. Autistics have no difficulty with this test, because their understanding of folk physics outstrips their understanding of folk psychology.

Folk physics is just part of a skill that Baron-Cohen calls 'systemising'. It is the ability to analyse input–output relationships in the natural, technical, abstract and even human world: to understand cause and effect, regularity and rules. He believes that human beings have two separate mental abilities, systemising and empathising, and that, though some people are good at both, others are good at one and bad at the other. Those who are good systemisers and bad empathisers will try to use their systemising skills to solve social problems. For instance, one person with Asperger's said to Baron-Cohen that 'Where do you live?' was not a good question, since it could be answered on many levels: country, city, district, street or house number. True, but most people solve the problem by empathising with the questioner. If speaking to a neighbour, he might name the house; if to a foreigner, the country.

If Asperger's people are good systemisers and bad empathisers, with extreme-male brains, the thought arises that there are probably people who are good empathisers and poor systemisers, with extreme female brains. A moment's thought will confirm that we all know such people, but their particular skill combination is rarely classified as pathological. It is probably easier to live a normal life in the modern world with poor systemising skills than with poor empathising skills. In the Stone Age, it might have been less easy.[38]

A MIND IN PARTS

The empathy story illustrates a very William James theme of separate instincts. To be good at empathising you need a domain, or module, in your mind that learns to intuitively treat animate creatures as having mental states as well as physical properties. To be good at systemising, you need a domain that learns how to intuit cause and effect, regularities and rules. These are separate mental modules, separate skills and separate learning tasks.

The empathy domain seems to rely on circuits around the paracingulate sulcus, a valley of the brain close to the mid-line and near the front of the head. In the studies of Chris and Uta Frith in London, this area lights up (in a suitable scanner) when a person reads a story that requires 'mentalising' – imagining the mental states of others; it does not light up when the person reads a story about physical cause and effect or a series of unlinked sentences. In people with Asperger's syndrome, however, this area does not light up when reading mental-state stories, but a neighbouring area does, instead. This is an area implicated in general reasoning, which supports the psychologists' hunch that Asperger's people reason about social issues rather than empathise about them.[39]

All of which rather supports the idea that Jamesian instincts must be manifest in mental circuits called modules, each specifically designed to be good at its specific mental task. Such a modular view of the mind was first enunciated by the philosopher Jerry Fodor in the early 1980s and later developed by the anthropologist John Tooby and the psychologist Leda Cosmides in the 1990s. Tooby and Cosmides were attacking the then widespread belief that the brain is a general-purpose learning device. Instead, said the anthropologist–psychologist couple, the mind is like a Swiss army knife. For blades and screw-drivers and things for helping Boy Scouts get stones out of horses' hoofs, read vision modules, language modules and empathy modules. Like the tools of a knife, these modules are rich in teleological purpose: it makes sense not just to describe what they are made of and how they do their job, but what they are for. Just as the stomach is for digestion, so the visual system of the brain is for seeing. Both are functional, and functional design implies evolution by natural selection, which implies at least partly a genetic ontology. The mind therefore consists of a collection of content-specific, information-processing modules adapted to past environments. Nativism was back.[40]

This was the high point of what is sometimes called the cognitive revolution. Though it now owes much to the tragic genius Alan Turing, with his extraordinary mathematical proof that reasoning could take a mechanical form – that it was a form of computation – the cognitive revolution really began with Noam Chomsky in the 1950s. Chomsky argued that the universal features of human language, invariant throughout the world, plus the logical impossibility of a child deducing the rules of a language as quickly as it does merely from the scanty examples available to it, must imply that there was something innate about language. Much later, Steven Pinker dissected the human 'language instinct', showed it had all the hallmarks of a Swiss army knife blade – structure designed for function – and added the notion that what the mind was equipped with was not innate data, but innate ways of processing data.[41]

Do not mistake this for an empty or obvious claim. It would be quite possible to imagine that vision, language and empathy are done by different parts of the brain in different people. That indeed is the logical prediction that follows from the empiricist argument that runs from Locke, Hume and Mill right up to the modern 'connectionists' who design multi-purpose computer networks to mimic brains. And it is wrong. Neurologists can produce battalions of case histories to support the idea that particular parts of the mind correspond to particular parts of the brain with very little variation all over the world. If you damage one part of your brain, in an accident or after a stroke, you do not suffer some generalised debility: you lose one particular feature of your mind – and the feature you lose depends precisely on which part of the brain is lost. This cannot but imply that different parts of the brain are pre-designed for different jobs, something that could only come about through genes. Genes are often thought of as constraints on the adaptability of human behaviour. The reverse is true. They do not constrain; they enable.

True, there have been rearguard actions by the retreating empiricists, but these skirmishes have only briefly delayed the advance of the modular mind. There is a degree of plasticity in the brain that allows different areas to compensate for the failure of their neighbouring area. Mriganka Sur has partly rewired the eyes of a ferret to the auditory cortex of its brain rather than the visual cortex, and in some rudimentary way it can still 'see', but not very well. Although you might think it remarkable that the ferret can see at all after such surgery, there is disagreement whether Sur's experiment reveals more about the plasticity of the brain or the limits of that plasticity.[42]

If the modular mind is real, then all you have to do to understand the special features of the human mind is to dissect the brain to find out which bits have 'hypertrophied' in the past few million years –

which modules and therefore which instincts are disproportionately big. Then you will know what makes human beings special. If only it were so easy! Almost everything in the human brain is bigger than in the chimpanzee brain. Human beings apparently do more seeing, more feeling, more moving, more balancing, more remembering and even more smelling than chimps. Far from finding a normal chimpanzee brain with a huge, turbo-charged thinking-and-speaking device attached to it, you find, if you look inside the human skull, more of everything. Closer inspection reveals that there are certain subtle disproportions. In primates generally, compared with rodents, the bits that do smelling have shrunk dramatically and the bits that do seeing have grown. The neocortex has grown at the expense of the rest. But even here the disproportion is not very marked. Indeed, since the neocortex develops last, and the frontal regions last of all, you could simply explain the big human brain as a chimp brain that has been grown for longer. In its extreme form this theory holds that the brain expanded, not because expansion was demanded by the requirement for it to do new functions – specifically language or culture – but because something required the enlargement of the brain stem itself and a bigger cortex came along as a passenger for the ride. Remember the lesson of the IQ domains in the ASPM gene: it is genetically easy just to make every part of the brain bigger. Once the big brain was there, hey presto, 50,000 years ago, *Homo sapiens* suddenly discovered he could use it to make bows and arrows, paint cave walls and think about the meaning of life.[43]

This idea has the advantage of again taking the species down a Cartesian peg – away goes the reassuring notion that humankind was the subject, rather than the object, in its own evolutionary story. But it is not necessarily incompatible with the idea of a modular mind. In fact, you could just as easily turn the logic on its head and argue that human beings were under selective pressure to develop more processing power in the parts of the brain needed for one function – language, say – and the easiest way for the genome to respond was to build a bigger brain generally. The ability to do more seeing and have a greater repertoire of moves was thrown in free. Besides, even a language module is hardly likely to be isolated from other functions. It needs fine discrimination of hearing, finer control of movement in the tongue, lips and chest, greater memory, and so on.[44]

Scientific theories, however, like empires, are at their most vulnerable when they have vanquished their rivals. No sooner had the modular mind triumphed than one of its main champions started dismantling it. In 2001 Jerry Fodor published a remarkable little book called *The Mind Doesn't Work That Way*, which argued that though breaking down the mind into separate computational modules was by far the best theory around, it did not and could not explain how the mind works.[45] Pointing out the 'scandalous' failure of engineers to build robots capable of routine tasks like cooking breakfast, Fodor gently reminded his colleagues how little had yet been discovered and chided Pinker for his cheerful optimism that the mind was explained.[46] Minds, said Fodor, are capable of abducting global inferences from the information supplied by the parts of the brain. You may see, feel and hear raindrops with three different brain modules linked to different senses, but somewhere in your brain resides the inference: 'it is raining'. In some inevitable sense, then, thinking is a general activity that integrates vision, language, empathy and other modules: mechanisms that operate as modules presuppose mechanisms that don't. And almost nothing is known about the mechanisms that are not modular. Fodor's conclusion was to remind scientists just how much ignorance they had discovered: they had merely thrown some light on how much dark there was.

But at least this much is clear. To build a brain with instinctive abilities, the Genome Organising Device lays down separate circuits with suitable internal patterns that allow them to carry out suitable computations, then links them with appropriate inputs from the senses. In the case of a digger wasp or a cuckoo, such modules may have to 'get the behaviour right' first time and may be comparatively indifferent to experience. But in the case of the human mind, almost all such instinctive modules are designed to be modified by experience. Some adapt continuously throughout life, some change rapidly with experience then set like cement. A few just develop to their own timetable. In the rest of this book, I propose to try to find the genes responsible for building – and changing – these circuits.

PLATONIC UTOPIA

One of the besetting sins of the nature–nurture debate has been the habit of utopianism, the notion that there is one ideal design of society that can be derived from a theory of human nature. Many of those who thought they understood human nature promptly turned description to prescription and set out a design of the perfect society. This practice is common to those on the nature side of the debate as well as those on the nurture side. Yet the only lesson to be drawn from utopian dreaming is that all utopias are hells. All attempts to design society by reference to one narrow conception of human nature, whether on paper or in the streets, end in producing something much worse. I propose to end each chapter mocking the utopia implied in taking any theory too far.

William James and the protagonists of instinct did not, as far as I can discern, write a utopia. But Plato's *Republic*, the father of all utopias, is in many ways close to a Jamesian dream. It is imbued with a similar nativism. The Republic has been called a 'managerial meritocracy' in which the same education is available to all, so the top jobs go to those with the innate talent for them.[47] In Plato's metaphorical republic (probably never intended as a political blueprint), everything is governed by strict rules. The Rulers, who make policy, are assisted by the Auxiliaries, who provide a sort of civil and defence service. Together these two classes are called the Guardians, and they are chosen on merit, which means on native talent. But to prevent corruption, the Guardians live lives of austere asceticism, unable to own property, to marry, or even to drink from gold cups. They live in a dormitory, but their miserable existence gladdens their hearts because they know it is for the good of the society as a whole.

Karl Popper was not the first, nor will he be the last, philosopher to call Plato's dream a totalitarian nightmare. Even Aristotle pointed out that there was not much point in a meritocracy if merit did not bring rewards – of wealth and sex as well as power: 'Men pay most attention to what is their own: they care less for what is common.'[48] Plato's citizens were expected to accept any spouse nominated by the state, and (if female) to suckle any baby. Some chance. But grant Plato the backhanded compliment of having this insight, at least: even the meritocracy is an imperfect society. If all people receive the same education, then the differences in their abilities will be innate. A truly equal-opportunity society merely rewards the talented with the best jobs and relegates the rest to doing the dirty work.

CHAPTER THREE

A convenient jingle

Professors are inclined to attribute the intelligence of their children to nature, and the intelligence of their students to nurture.

Roger Masters[1]

Disagreement thrives on uncertainty. In the 1860s, uncertainty over the source of the Nile was the source of a bitter dispute between two English explores, John Hanning Speke and Richard Burton. Only two men who have shared a camp for many months could disagree so violently. Speke favoured Lake Victoria, which he had discovered while Burton lay ill in a tent at Tabora; Burton insisted that the source lay in or near Lake Tanganyika. The feud only ended in 1864 when Speke shot himself (perhaps accidentally) on the day he was to debate with Burton in public. Speke, by the way, was right.

Watching this dispute from an influential position in the Royal Geographical Society, and occasionally fanning the flames on behalf of Burton, was a distinguished geographer by the name of Francis Galton. It was Galton's fate to ignite an even bigger feud in that same year, one that would run for more than a century: nature versus nurture. The nature–nurture debate is a bit like the argument over the source of the Nile. Both thrived on ignorance; the more that came to be known, the less the argument seemed to matter. Both seemed unnecessarily petty. Surely, what mattered more than which lake was the source of the Nile was that Africa contained two vast lakes new to Western science. Likewise, it surely matters less whether human nature is more innate or more learned, but instead the precise way in which it is both. The Nile is the sum of thousands of streams, no one of which can be truly called its source; the same is true of human nature.

Galton's passion was quantifying. In a long career, he invented, coined or discovered a wide range of things: northern Namibia, anticyclone weather systems, the study of twins, questionnaires, fingerprints, composite photographs, statistical regression and eugenics. But perhaps his most lasting legacy is to have inaugurated the nature–nurture debate and coined the very phrase. Born in 1822, he was a grandson of the great scientist, poet and inventor, Erasmus Darwin, by his second wife. He found his half-cousin Charles's theory of natural selection both convincing and inspiring, ascribing this immodestly to 'an hereditary bent of mind that both its illustrious author and myself have inherited from our common grandfather, Dr Erasmus Darwin'. Thus emboldened by his own pedigree, he now found his true calling in the statistics of heredity. In 1865, deserting geography, he published an article on 'hereditary talent and character' in *Macmillan's Magazine*, in which he revealed that distinguished men had distinguished relatives. He expanded it into a book called *Hereditary Genius* in 1869.

Galton was simply asserting that talent runs in families. Exhaustively and enthusiastically, he described the pedigrees of famous judges, statesmen, peers, commanders, scientists, poets, musicians, painters, divines, oarsmen and wrestlers. 'The arguments by which I endeavour to prove that genius is hereditary, consist in showing how large is the number of instances in which men who are more or less illustrious have eminent kinsfolk.'[2] It was not very sophisticated reasoning. After all, one might just as well argue the opposite, that the rise of humble men to great eminence would reveal their innate talents triumphing over the disadvantages of circumstance; the clustering of talent in families might indicate shared teaching. Most reviewers thought Galton had overstated the role of heredity and ignored the contribution of upbringing and family. In 1872 a Swiss botanist, Alphonse de Candolle, asserted as much at book length. Candolle pointed out that great scientists in the previous two centuries had come from countries or cities with religious tolerance, widespread trade links, a moderate climate and democratic governments – suggesting that achievement owed more to circumstance and opportunity than to native genius.[3]

Candolle's attack stung Galton into a second book, *English Men of Science: Their Nature and Nurture*, in 1874, in which he employed a questionnaire for the first time, and repeated his conclusion that scientific geniuses were born, not made. It was in this book that he coined the famous alliteration:

> The phrase 'nature and nurture' is a convenient jingle of words, for it separates under two distinct heads the innumerable elements of which personality is composed.[4]

He may have borrowed the phrase from Shakespeare, who in *The Tempest* has Prospero insult Caliban thus:

> A devil, a born devil, on whose nature nurture can never stick.[5]

Shakespeare was not the first to juxtapose the two words. Three decades before *The Tempest* was first performed, an Elizabethan schoolmaster by the name of Richard Mulcaster, the first headmaster of the Merchant Taylors' school, was so fond of the antiphony of nature and nurture that he used it four times in his 1581 book *Positions Concerning the Training Up of Children*:

> ... [Parents] will have their children nursed as well as they can, without question where, or quarrelling by whom: so as they may have that well brought up by nurture, which they love so well, bequeathed them by nature ... God hath provided that strength in nature, wherby he entendes no

exception in nurture, for that which is in nature . . . Which naturall abilities, if they be not perceived, by whom they should: do condemne all such, either of ignoraunce, if they could not judge, or of negligence, if they would not seeke, what were in children, by nature emplanted, for nurture to enlarge . . . Which being thus, as both the truth tells the ignorant, and reading shewes the learned, we do wel then perceave by naturall men, and Philosophicall reasons, that young maidens deserve the traine: bycause they have that treasure, which belongeth unto it, bestowed on them by nature, to be bettered in them by nurture.[6]

He repeated the phrase in his next book *Elementaries* in 1582: 'whereto nature makes him toward, but that nurture sets him forward'. Mulcaster was a curious character. Born in Carlisle, he was a distinguished scholar and famous, if strict, educational reformer. He quarrelled irascibly with the school governors and was a passionate advocate of the game of football: 'The foteball strengtheneth and brawneth the whole body,' he observed. Mulcaster also dabbled in drama, writing several pageants for the royal court, and educating the playwrights Thomas Kyd and Thomas Lodge at his school. He is supposed by some to have been the model for the character of Holofernes, the vain schoolmaster in *Love's Labour's Lost*, so there is a good chance that Shakespeare either knew Mulcaster or read him.

Shakespeare may also have been the inspiration for the next of Galton's ideas. Two of Shakespeare's plays turn on the confusion of twins: *The Comedy of Errors* and *Twelfth Night*. Shakespeare was himself the father of twins, and he used mistaken twins to make fiendishly ingenious plots. But, as Galton pointed out, in *A Midsummer Night's Dream* Shakespeare introduced a pair of 'virtual twins' – unrelated individuals who had been reared together. Hermia and Helena, despite being 'like to a double cherry, seeming parted, but yet an union in partition',[7] not only look physically unlike each other, but are attracted to different men and end up quarrelling violently.

Galton followed up the hint. The next year he wrote an article entitled 'The history of twins, as a criterion of the relative powers of nature and nurture'. At last he had a respectable way to test the heredity hypothesis, free of the objections raised against his pedigrees. Remarkably, he deduced that there were two sorts of twin: identical twins, born from 'two germinal spots in the same ovum', and non-identical twins 'each from a separate ovum'. This is not bad. For 'germinal spot' read nucleus and you are close to the truth. Yet in both kinds, the twins shared nurture. So if identical twins resembled each other in behaviour more than fraternal twins, then the influence of heredity was supported.

Galton wrote to 35 pairs of identical twins and 23 pairs of non-identical twins, collecting anecdotes of their similarity and difference. Triumphantly he recounted the results. Twins that resembled each other from birth remained similar throughout their lives, not only in appearance but also in ailments, personality and interests. One pair suffered severe toothache in the same tooth at the same age. Another pair bought identical sets of champagne glasses as presents for each other at the same time at different ends of the country. Twins that were born different, by contrast, grew more different as they grew older. 'They were never alike either in body or mind, and their dissimilarity increases daily,' said one of his respondents. 'The external influences have been identical; they have never been separated.' Galton sounded almost embarrassed by the strength of his conclusion: 'There is no escape from the conclusion that nature prevails enormously over nurture ... My fear is, that my evidence may seem to prove too much, and be discredited on that account, as it appears contrary to all experience that nurture should go for so little.[8]

SPLITTING PAIRS

With hindsight one can pick all sorts of holes in Galton's first twin study. It was anecdotal, small, and the argument was circular: twins that appeared identical behaved identically. He had not distinguished identicals from fraternals genetically. Yet the study was remarkably persuasive. By the end of his life Galton had seen his hereditarian beliefs move from scepticism to orthodoxy. 'Nature limits the powers of the mind as definitely as those of the body,' said *The Nation* in 1892, 'On these points, among thinkers everywhere, [Galton's] opinions have prevailed.'[9] The old empiricism of John Locke, David Hume and John Stuart Mill, whereby the mind was seen as a blank sheet of paper on which experience would write its script, had been replaced by a sort of neo-Calvinist notion of inherited individual destiny.

There are two ways to look at this development. You can damn Galton for being seduced by his 'convenient jingle' into presenting a false dichotomy. You can see him as one of the evil spirits of the twentieth century, cursing the three generations that followed to swinging like a pendulum between ridiculous extremes of environmental and genetic determinism. You can note with horror that from the beginning, Galton's motives were eugenic. On the very first page of *Hereditary Genius* in 1869 he was already extolling the virtues of 'judicious marriage', lamenting the 'degradation of human nature' by the propagation of the unfit and invoking the 'duty' of the authorities to exercise power to change human nature by progressive breeding. These suggestions would grow into the pseudo-science of eugenics. With hindsight, therefore, you can blame him for an idea that would cause misery and cruelty to millions in the century to come, not just in Nazi Germany but in some of the most tolerant countries of the world.[10]

All this would be true, though it is a little harsh to expect that none of it would have happened without Galton, let alone that he should have foreseen where his ideas would lead. Even the convenient jingle would have soon occurred to somebody else. A more charitable reading of history would see Galton as a man far ahead of his time who hit upon a remarkable truth: that many aspects of our behaviour start within us in some way, that we are not putty in the hands of society or victims of our surroundings. You could even – though this might be stretching it – assert that this notion was vital in keeping alive the flame of liberty in the environmentalist despotisms of the twentieth century: those of Lenin, Mao and their imitators. Galton's insights into heredity were remarkable, considering he knew nothing about genes.

He would have had to wait more than a century to see that the study of twins did in the end prove much of what he had suspected. To the extent that they can be teased apart, nature prevails over one kind of (shared) nurture when it comes to defining *differences* in personality, intelligence and health between people *within the same society*. Note the caveats.

This is a recent development. Twenty years ago, the picture was very different. By the 1970s the whole notion of studying twins to learn about heredity was in eclipse. Two of the largest studies of twins since Galton were in disgrace. In Auschwitz, Josef Mengele was notoriously fascinated by twins. He sought them out among new arrivals at the concentration camp, and segregated them into special quarters for study. Ironically, this 'favouritism' led to a higher survival rate among twins than singletons – most of the small children who survived Auschwitz were twins. In exchange for submitting to procedures that were often brutal and sometimes fatal, they were at least better fed. All the same, few survived.[11]

In Britain meanwhile, the educational psychologist Cyril Burt was slowly accumulating a set of identical twins reared apart, which enabled him to calculate the heredity of intelligence. In 1966, when he published the full set of results, he claimed to have found 53 pairs of such twins. This was an extraordinarily large sample, and Burt's conclusion that IQ was highly heritable influenced British education policy. But it later emerged that at least some of the data was almost certainly faked. The psychologist Leon Kamin noticed that the correlation had remained exactly the same, to the third decimal place, even while the data set had expanded over several decades. The *Sunday Times* simultaneously asserted that two of Burt's co-authors probably did not exist (one has since reappeared, however).[12]

With a history like this, it was little wonder that twin research was a tainted subject in the 1970s. Yet today the study of twins is reborn as the principal method of a scientific discipline known as behaviour genetics that has flowered especially in the United States, Holland, Denmark, Sweden and Australia. It is sophisticated, argumentative, mathematical and expensive – everything that a thoroughly modern science should be. But at its core lies Galton's insight: that human twinning provides a beautiful natural experiment for discerning the contributions of nature and nurture.

In this respect, fortune has been generous to human beings. The ability to produce identical twins seems to be rare in the animal kingdom. It is unknown in mice, for example, which produce litters of non-identical litter-mates. Human beings occasionally produce litters, too. Among white people, about one birth in every 125 consists of two non-identical, fraternal or 'dizygotic' twins – derived from two zygotes or fertilised eggs. The rate is higher among Africans and lower among Asians. But one birth in every 250 consists of identical (or monozygotic) twins, derived from a single fertilised egg. Without a genetic test, identical twins cannot be reliably distinguished from fraternal twins, though there are telltale signs. Their ears tend to be identical.[13]

Behaviour genetics is a simple matter of measuring how similar are identical twins, how different are fraternals, and how both identicals and fraternals turn out if separately adopted into different families. The result is an estimate of 'heritability' for any trait. Heritability is a slippery concept, much misunderstood. For a start, it is a population average, meaningless for any individual person: you cannot say that Hermia has more heritable intelligence than Helena. When somebody says that the heritability of height is 90 per cent, he does not and cannot mean that 90 per cent of my inches came from my genes and 10 per cent from my food. He means that the variation in height *in a particular sample* is attributable 90 per cent to genes and 10 per cent to environment. There is no variability in height for the individual and therefore no heritability.

Moreover, heritability can only measure variation, not absolutes. Most people are born with ten fingers. Those with fewer have usually lost some through accidents – through the effects of the environment. The heritability for finger number is therefore close to zero. Yet it would be absurd to argue that environment is the cause of us having ten fingers. We grow ten fingers because we are genetically programmed to grow ten fingers. It is the variation in finger number that is environmentally determined; the fact that we have ten fingers is genetic. Paradoxically, therefore, the least heritable features of human nature may be the most genetically determined.[14]

So, too, with intelligence. It cannot be right to say that Hermia's intelligence is caused by her genes: it is obvious that you cannot become intelligent without food, parental care, teaching or books. Yet in a sample of people who have all these advantages, the variation between who does well in exams and who does not could indeed be down to genes. In that sense, variation in intelligence can be genetic.

Through accident of geography, class or money, most schools have pupils from similar backgrounds. By definition, they give them similar teaching. Having therefore minimised the differences in environmental influences, they have unconsciously maximised the role of heredity: it is inevitable that the difference between the high-scoring and the low-scoring pupils must be down to their genes, for that is just about all that is left to vary. Again, heritability is a measure of what is varying, not what is determining.

Likewise, in a true meritocracy, where all have equal opportunity and equal training, the best athletes will be the ones with the best genes. Heritability of athletic ability will approach 100 per cent. In the opposite kind of society, where only the privileged few get sufficient food and the chance to train, background and opportunity will determine who wins the races. Heritability will be zero. Paradoxically, therefore, the more equal we make society, the higher heritability will be, and the more genes will matter.

COINCIDENCE

I've laboured the caveats deliberately before even mentioning the results of modern twin studies. The story of those studies begins in 1979, when there appeared in a Minneapolis newspaper an account of a pair of identical twin men from western Ohio reunited at the age of 40. Jim Springer and Jim Lewis had been reared apart in adopted families since they were a few weeks old. Intrigued, the psychologist

Thomas Bouchard asked to meet them to record their similarities and differences. Within a month of their re-encounter, Bouchard and his colleagues examined the Jim twins for a day and were astonished by the similarities. Though they had different hairstyles, their faces and voices were almost indistinguishable. Their medical histories were very similar: high blood pressure, haemorrhoids, migraines, 'lazy eye', chain-smoked Salem cigarettes, bitten nails, weight gain at the same age. As expected, their bodies showed remarkable similarity. But so did their minds. Both followed stock-car racing and disliked baseball. Both had carpentry workshops. Both had built a white seat around a tree trunk in the garden. They went to the same Florida beach on vacation. Some of the coincidences were, well, coincidences. Both had dogs named Toy. Both had wives named Betty. Both had divorced women named Linda. Both had named their first children James Alan (though one spelled it James Allen).

It occurred to Bouchard that maybe twins reared apart would turn out to be not just as similar, but maybe more similar than twins reared together. In the same family, differences might become exaggerated: one twin would start to do a little more of the talking and the other less, or something like that. This is now known to be true. Twins, like the Jims, who were separated early in life, have more similarities than twins separated at a later age.

The news reporter who had first written about the Jim twins interviewed Bouchard after his meeting with them, and the resulting article brought a flood of interest from the media. The Jim twins appeared on the *Tonight* show, with Johnny Carson, and that was when things began to snowball. Twins started calling. Bouchard invited them to Minnesota and put them through a battery of physical and psychological tests, administered eventually by a team of 18 people. By the end of 1979, 12 pairs of reunited twins had contacted Bouchard. By the end of 1980, 21; a year after that, he had 39 pairs.[15]

That was the year Susan Farber published a book definitively rubbishing all studies of identical twins reared apart as unreliable.[16] The studies exaggerated similarities, ignored differences and skated over the fact that many twins had spent many months together as infants before their adoption or had been reunited many months before being seen by scientists. Some of the studies, such as Cyril Burt's, were perhaps even fabricated altogether. Farber's book was seen as the last word on the matter, but Bouchard merely saw it as a challenge to do a flaw-free study. He was determined not to leave himself open to such accusations, and he carefully recorded everything about his twin pairs. Anecdotes aside, he was gathering real, quantitative information on similarity. By the time he published, his data were all but impregnable to the Farber criticisms. Not that this impressed the establishment. His critics still charged that he was proving nothing but his own assumptions. Of course these people resembled each other – they lived in similar middle-class suburbs of similar cities; they swam in the same cultural sea; they were taught the same Western values.

All right, then, said Bouchard, and he set out to find fraternal (dizygotic) twins reared apart. These were people who had shared a womb as well as a Western upbringing. If his critics were right, then they too should show remarkable similarities of mind.[17] Do they?

Take religious fundamentalism. In a recent study Bouchard measured how fundamentalist individuals are by giving them questionnaires about their beliefs. The correlation between the resulting scores for identical twins reared apart is 62 per cent; for fraternal twins reared apart it is just 2 per cent. Bouchard repeats the exercise with a different questionnaire designed to elicit a broader measure of religiosity and still gets a strong result: 58 per cent versus 27 per cent. He then shows a similar contrast between sets of identical twins reared together and fraternal twins reared together. He repeats the exercise with a different questionnaire designed to discover what he calls 'right wing attitudes'. Again there is a high correlation in identical twins reared apart (69 per cent) and no correlation at all in fraternal twins reared apart. He gives the twins a different questionnaire that simply lists single phrases and asks for approval or disapproval: immigrants, death penalty, X-rated movies, etc. Those who reply no to immigrants, yes to the death penalty, and so on are judged more 'right wing'. The identical-apart correlation is 62 per cent, the fraternal-apart correlation only 21 per cent. Similar huge differences emerge from similar large studies in Australia.[18]

Bouchard is not trying to prove that there is a God gene or an anti-abortion gene. Nor is he trying to claim that the environment plays no part in determining details of religious observance. It is absurd to argue, for instance, that the reason Italians are Catholic and Libyans are Muslim is because they possess different genes. He is simply claiming that astonishingly, even in such a prototypically 'cultural' thing as religion, the impact of genes cannot be ignored and can be measured. There is a partly heritable aspect of human nature, which might be called religiosity, and it is distinct from other attributes of personality (it correlates poorly with other measures of personality such as extroversion). This can be detected using simple questionnaires and it predicts fairly well who will end up becoming a fundamentalist believer within any particular society.

Notice how even this one simple study refutes many of the objections raised by critics of behaviour genetics. Many people argue that questionnaires are unreliable and crude measures of people's real thoughts; but that simply makes these results conservative. The effects would probably be bigger if measurement error could be ruled out. Many argue that identical twins reared apart have not really lived such separate lives as is claimed. They have often been reunited for many years before the experiment is done. But if this is true, it will be just as true for the fraternal twins reared apart. The same response demolishes the frequent objection that Bouchard, by attracting self-selected twins to his studies, preferentially attracts those who are more similar to each other.[19] But it is the differences between identical and fraternal twins that are revealing, not the absolute similarity. Others say you cannot separate nature from nurture because they interact. True, but the fact that twins apart do not differ greatly from twins together suggests that such an interaction is less powerful than many believe.

In researching this book, I encountered a vitriolic opinion of Bouchard's research among many people. Not content with making the long-since-answered arguments in the last paragraph, they would pointedly remind me to check where Bouchard got the funds for his research: the Pioneer Fund. This fund, founded in 1937 by a textile billionaire, is unashamedly in favour of eugenics. Its charter reads: 'To conduct or aid in conducting study and research into the problems of heredity and eugenics in the human race generally and such study and such research in respect to animals and plants as may throw light upon heredity in man, and research and study into the problems of human race betterment with special reference to the people of the United States.'[20] Based in New York, it is run by a board consisting mainly of ageing war heroes and lawyers.

Their motive in supporting Bouchard's research is presumably that they want to believe that genes influence behaviour, so they give money to a researcher who seems to be getting results that support such a conclusion. Does this mean Bouchard and all his many colleagues (not to mention the similar twin-studiers in Virginia, Australia, Holland, Sweden and Britain) have faked their data to please their funders? Seems pretty far-fetched. Besides, you only have to meet Bouchard for a few minutes to realise that he is nobody's patsy and nobody's fool, let alone a raving determinist itching to unleash a new eugenics movement on the world. He takes the Pioneer Fund money because it has no strings attached. 'My rule is that if they don't make any restrictions on me – what I think, what I write, what I do – I'll accept their money.'[21]

There is, of course, a problem with how such studies are reported. The headline – 'the gene for x' – does much mischief, not least because of the reputation genes have garnered for being invincible bulldozers of all that stands in their path. However, the champions of nurture must bear some responsibility for creating this reputation in the first place, by equating genes with inevitability in the process of arguing that since behaviour is not inevitable, so genes cannot be involved. Nurture champions repeatedly state that 'the gene for x' means a gene that always and only causes that behaviour; nature champions reply that they merely mean that the gene increases the probability of behaviour x, compared with other versions of the same gene.[22] When the British twin-researcher Thalia Eley announced in 1999 that evidence from 1,500 identical versus fraternal twin pairs in Britain and Sweden suggested a strong genetic influence on whether an individual child would become a school bully, should she have complained or apologised when a reporter described her conclusion in the usual shorthand: 'bullying behaviour may be genetic'?[23] The truer statement would be 'variations in bullying behaviour may be genetic in typical Western societies', but few reporters can expect news editors to insert such caveats.

It is worth recalling how much of a shock the carefully controlled twin studies of the 1980s were when they first came out. Until then it was genuinely thought that differences in experience even among middle-class Westerners would produce differences in personality with no help from the genes. The hypothesis on trial was not 'all in the genes', it was 'not in the genes at all'. Here is a quote from a leading textbook of personality psychology, published in 1981, the year Bouchard first had good data: 'Imagine the enormous differences that would be found in personalities of twins with identical genetic endowments if they were raised in two different families.'[24] That is what everybody thought, even Bouchard. 'Look,' he says openly, 'when I started, I did not believe these kinds of things could be influenced by genes. I was persuaded by the evidence.'[25] The twin studies have caused a genuine revolution in the understanding of personality.

However, the very success of behaviour genetics has been its undoing. Its results are boringly predictable: everything turns out to be heritable. Far from being able to parcel the world into genetic and environmental causes, as Galton wanted, twin studies have found almost everything to be equally strongly heritable. When Bouchard began, he expected to find that some measures of personality were more heritable than others. But at the end of two decades of such separated-twin studies in many countries with larger and larger samples of twins, there is an unambiguous conclusion. For nearly all measures of personality, heritability is high in Western society: identical twins raised apart are much more similar than fraternal twins raised apart.[26] The difference between one individual and another owes more to differences in their genes than factors in their family background.

Psychologists nowadays define personality in five dimensions- the so-called 'big five' factors: Openness, Conscientiousness, Extroversion, Agreeableness and Neuroticism (OCEAN for short). Questionnaires can elicit personal scores for each of these dimensions and they seem to vary independently. You can be open-minded (O), fussy (C), extrovert (E), jealous (A) and calm (N). In each case a little over 40 per cent of the variation in personality is due to direct genetic factors, less than 10 per cent due to shared environmental influences (i.e., mostly the family), and about 25 per cent due to unique environmental influences experienced by the individual (everything from illness and accident to the company he or she keeps at school). The remaining 25 per cent or so is simply measurement error.[27]

In a sense what these twin studies have proved is that the word 'personality' means something. When you describe somebody as having a certain personality, you are intending to refer to some intrinsic part of their nature that is beyond the influence of other people – the content of their character, to borrow a famous phrase. By definition, you mean something unique to them. It is, however, counter-intuitive after a century of Freudian certainties to find how little that intrinsic character is influenced by the family they grew up in.[28]

In this respect, personality is about as heritable as body weight. The correlation between two siblings in weight, according to one study, is 34 per cent. The similarity between parents and children is a little lower, at 26 per cent. How much of this similarity is due to the fact that they live together and eat similar food, and how much to the fact that they share many of the same genes? Well, identical twins reared in the same family have a correlation of 80 per cent while fraternal twins reared together have only 43 per cent similarity, which suggests that genes matter rather more than shared eating habits. What about adoptees? The correlation between adoptees and their adoptive parents is only 4 per cent, and between unrelated siblings in the same family it is just 1 per cent. By contrast, identical twins reared apart in different families are still 72 per cent similar in weight.[29]

Conclusion: weight is largely due to genes, not eating habits, so throw away the diet advice and let rip with the ice cream? Of course not. The study says nothing about the causes of weight; it only reveals something about the causes of differences in weight within a particular family. Given the same access to food, some people will put on more weight than others. People are getting fatter in Western societies, not because their genes are changing, but because they are eating more and taking less exercise. But with everybody having similar access to food, the ones who put on weight fastest will be the ones with certain genes. So variation in weight can be inherited, even while changes in the average can be environmental.

What kind of gene could cause personality to vary? A gene is a set of instructions for making a protein molecule. To leap from this epitome of digital simplicity to the complexity of personality sounds impossible. Yet it can now, for the first time, be done. The changes in genetic sequence that lead to changes in character are being found: the haystack is revealing its first few needles. Take the gene for a protein called brain-derived neurotrophic factor, or BDNF, on chromosome 11. It is a short gene, a chunk of DNA text just 1,335 letters long – exactly the same length as this paragraph, by good fortune. The gene spells out in four-letter code the complete recipe for a protein that acts as a sort of fertiliser in the brain encouraging the growth of neurons, and probably does much else besides. In most animals, the 192nd letter in the gene is G, but in some people it is A. About three-quarters of human genes carry the G version, the rest the A version. This minuscule difference, just one letter in a long paragraph, causes a slightly different protein to be built – with methionine instead of valine at the 66th position in the protein. Since everybody has two copies of each gene, that means there are three kinds of people in the world: those with two methionines in their BDNFs, those with two valines and those with one of each. If you give people a questionnaire about their personality and simultaneously find out which kind of BDNF they have, you will find a striking effect. The met–mets are noticeably less neurotic than the val–mets, who are noticeably less neurotic than the val–vals.[30]

The val–vals are the most, and the met–mets the least, depressed, self-conscious, anxious and vulnerable – four of the six facets that make up the psychologists' dimension of neuroticism. Of the other 12 facets of personality, only one (openness of feelings) shows any association. This gene, in other words, specifically affects neuroticism.

Do not get carried away. This finding accounts for only a small portion of the variation between people, perhaps 4 per cent. It may prove to be a peculiarity of 257 families in Tecumseh, Michigan, where the study was done. It is most definitely not 'the' neuroticism gene. But at least in Tecumseh it is a gene whose variation explains some of the personality differences between any two individuals and in a way that is consistent with the standard way of describing personality. It is also the first gene to associate so strongly with depression, which gives a faint glimmer of medical hope for one of the least treatable and commonest disorders of modern life. The lesson I wish to draw from it is not that this particular gene will prove especially significant, but that it proves just how easy it is to make the leap from a spelling change in a DNA code to a real difference in personality. Neither I nor anybody else can yet begin to tell you how or why such a tiny change results in a different personality, but that it does so seems all but certain. The appeal to incredulity beloved by some of the critics of behaviour genetics – 'genes are just recipes for proteins, not determinants of personality' – just will not wash. A change in a protein recipe can indeed result in a change in personality. There are other candidate genes emerging, too.

So it is not so crazy to conclude that people differ in personality more if they have different genes than if they are reared in different families. Hermia is less like Helena, despite being raised together, than Sebastian is like Viola even though they were raised apart. This might seem obvious to the point of banality. Any parent who has more than one child notices drastic differences in personality and knows for sure that he did not put them there. But then parents are almost bound to notice the innate differences, because parents are holding the environment fairly constant by raising each child in the same family. The surprise of the twins-apart studies is that they seem to show that, even when the environments are varied somewhat, the differences in personality are still mostly innate. Even when the family environment does vary, it leaves no mark on personality. This conclusion emerges most starkly from the study of twins, but it is fully supported by other studies of adoption and of the relations of twins and adoptees.

The effect of being reared in the same home is negligible for many psychological traits.[31]

Or:

The shared environment plays only a small and non-significant role in the creation of personality differences in adults.[32]

Quickly but imperceptibly, statements like this seem to evolve into the assertion that families do not matter. Go ahead, neglect your kids, the logic seems to follow; their personality will not be affected. Some blame the researchers themselves for leaving this impression. Read the small print, however, and you will always find careful denials of such a fallacy. A happy family gives you other things than personality – things like happiness. Families do matter for personality; a child desperately needs to be reared in a family in order to develop her personality. So long as she does have a family to grow up within, it does not terribly matter whether the family is big or small, rich or poor, gregarious or solitary, old or young. A family is a bit like vitamin C: you need it or you will become ill, but once you have it, consuming extra does not make you healthier.

For those attached to the idea of the meritocracy, this is an encouraging discovery. It means there is no excuse to discriminate against those from underprivileged backgrounds, or be wary of those brought up in unusual families. A disadvantaged childhood does not condemn a person to a certain personality. Environmental determinism is at least as heartless a creed as genetic determinism, a theme I shall have cause to revisit throughout this book. So it is lucky we do not have to believe in either.

There is a criticism to be made of twin studies of personality, one that I shall weave into my argument that genes are the agents of nurture at least as much as they are the agents of nature. The criticism rests on the fact that heritability depends entirely on context. The heritability of personality may be high in a group of middle-class Americans who have experienced equivalent, nay identical, patterns of nurture. But throw a few orphans from Sudan or the offspring of New Guinea headhunters into the sample and the heritability of personality would probably drop rather fast: now environment would matter. Hold the environment constant and it's the genes that vary: what a surprise! 'I can prove in a court of law,' says Tim Tully, who studies the genes of memory but has no time for twin studies, 'that heritability has nothing to do with biology.'[33] To the extent, therefore, that twin-studiers try to suggest that the measurement of heritability is an end in itself, they are deluding themselves. And having once produced surprisingly strong evidence that genes do affect personality, it is not clear what they go on to do. Twin studies alone are notoriously unhelpful at revealing which actual genes are involved.

Here's why. Heritability is usually highest for those features of human nature caused by lots of genes rather than by the action of single genes. And the more genes are involved, the more the heritability is actually caused by the side effects of genes rather than the direct effect. Criminality, for instance, is pretty highly heritable: adopted children end up with a criminal record that looks a lot more like that of their biological parents than that of their foster parents. Why? Not because there are specific criminality genes, but because there are specific personalities who get into trouble with the law and those personalities are heritable. As Eric Turkheimer, a twin-studier, puts it, 'Does anyone really suppose that unintelligent, unattractive, greedy, impulsive, emotionally unstable, or alcoholic people are no more likely than anyone else to become criminals or that any of these characteristics could be completely independent of genetic endowment?'[34]

INTELLIGENCE

Despite the sweeping successes of twin studies, a few features of human behaviour prove to be less heritable. The sense of humour shows low heritability: adopted siblings seem to have quite similar senses of humour, while separated twins have rather different ones. People's food preferences seem to be barely heritable – you get your food preferences from your early experience, not your genes (so do rats).[35] Social and political attitudes show a strong influence from the shared environment – liberal or conservative parents seem to be able to pass on their preferences to their children. Religious affiliation, too, is passed on culturally, rather than genetically, though not religious fervour.

What about intelligence? The debate about the heritability of IQ has been scarred by controversy since its inception. The first IQ tests were crude and culturally biased. In the 1920s, convinced that intelligence was largely hereditary and alarmed at the thought of excessive breeding by stupid people, governments in the United States and many European countries began to sterilise mental defectives to prevent them passing on their genes. Then in the 1960s came a sudden revolution, as in so many other debates. From now on, even the assertion of heritable IQ led to vitriolic campaigns of denunciation, assaults on your reputation and demands for your dismissal. The first to suffer the treatment was Arthur Jensen in 1969 following his article in the *Harvard Educational Review*.[36] By the 1990s, the argument that society was segregating itself by assortative mating along intellectual and therefore racial lines – asserted in *The Bell Curve* by Richard Herrnstein and Charles Murray – provoked another fury of rage among academics and journalists.[37]

Yet I suspect if you took a poll of ordinary people, they would hardly have changed their views over a century. Most people believe in 'intelligence' – a natural aptitude or lack of it for intellectual pursuits. The more children they have, the more they believe in it. This does not stop them also believing in coaxing it out of the gifted and coaching it into the ungifted through education. But they think that there is something innate.

The twin studies, reared apart or together, unambiguously support the idea that despite the fact that some people are good at some things and others good at other things, there is such a thing as unitary intelligence. That is to say, most measures of intelligence correlate with each other. People who are good at general knowledge tests or vocabulary tests are usually good at abstract reasoning or number-series completion tasks. This was first noticed a century ago by Galton's follower, the statistician Charles Spearman, who dubbed the common factor 'g' for general intelligence. Today, a measure of g derived from correlating different IQ tests remains a powerful predictor of how well a child will do at school. There has been more research on g than on any other subject in psychology. Theories of multiple intelligence come and go, but the notion of correlated intelligence just will not go away.

What is g? Something that appears so real in statistical tests must surely have a physical manifestation in the brain. Is it something to do with speed of thought, or size of brain, or is it something subtler? The first thing to be said is that the search for the genes of g has been a huge disappointment. None of the genes that are capable of causing mental retardation when broken prove to have any effect on intelligence when altered more subtly. Searching at random through the genes of intelligent people to find ways in which they consistently differ from normal people has so far turned up just one decent statistical correlation (for the IGF2R gene on chromosome 6) and more than 2,000 no-shows. This may just mean that the haystack is too big and needles too small. Candidate genes, such as the PLP gene that seems to affect speed of neuronal transmission, have proved capable of explaining only a small degree of reaction time and do not correlate well with g: the speedy-brain theory of intelligence does not look promising.[38]

The one physical feature that does clearly predict intelligence is brain size. The correlation between brain volume and IQ is about 40 per cent, a number that leaves much room for the small-brained genius and the big-brained dullard, but is still a strong correlation.

Brains are composed of white matter and grey matter. When, in 2001, brain scanners reached the stage that people could be compared for the amount of grey matter in their brains, two separate studies in Holland and Finland found a high correlation between g and volume of grey matter, most especially in certain parts of the brain. Both also found a huge correlation between identical twins in grey matter volume: 95 per cent. Fraternal twins had only 50 per cent correlation. These figures indicate something that is under almost pure genetic control, leaving very little room for environmental influence. Grey matter volume must be 'due completely to genetic factors and not to environmental factors' in the words of Danielle Posthuma, the Dutch researcher. These studies bring us no closer to the actual genes of intelligence, but they leave little doubt that they are there. Grey matter consists of the bodies of neurons, and the new correlation implies that clever people may literally have more neurons, or more connections between neurons, than do normal people. After the discovery of the ASPM gene's role in determining brain size through neuron number (chapter 1), it is beginning to look as if some of the genes of g will soon be found.[39]

However, g is not everything. Twin studies of intelligence also reveal a role for the environment. Unlike personality, intelligence does seem to receive a strong influence from the family. Studies of the heritability of IQ in twins, adoptees and combinations of the two have all gradually converged on the same conclusion. IQ is approximately 50 per cent 'additively genetic', 25 per cent influenced by the shared environment and 25 per cent influenced by environmental factors unique to the individual. Intelligence therefore stands out from personality in being much more susceptible to family influence. Living in an intellectual home does make you more likely to become an intellectual.

However, these average figures conceal two very much more interesting features. First, you can find samples of people in which variation in IQ is much more environmental and much less genetic than the average. Eric Turkheimer found that the heritability of IQ depends strongly on socio-economic status. In a sample of 350 twin pairs, many of which had been raised in extreme poverty, there emerged a clear difference between the richest and the poorest. Among the poorest children, practically all the variability between individual IQ scores was accounted for by shared environment and none by genetic type; in the richer families, the opposite was true. In other words, living on a few thousand dollars a year can severely affect your intelligence for the worse. But living on $40,000 a year or $400,000 a year makes little difference.[40]

This is a finding with obvious policy significance. It implies that raising the safety net of the poorest does more to equalise opportunity than reducing inequality in the middle classes. It is dramatic confirmation of the truth I alluded to earlier: that even when variation in achievement is explained entirely by genes, this does not mean the environment does not matter. The reason you find such strong genetic effects in most samples is because most people live in adequately happy, supportive and affluent families. If they did not, they would suffer enormously. It is a point that is almost certainly true of personality, too. Your parents may not have been able to alter your adult personality by being a little bit strict. But you can be sure that they would have done so if they had locked you in your room ten hours a day for weeks on end.

Recall the heritability of weight. In a Western society, with ample access to food, those who put on weight faster will be the ones with the genes that nudge them into eating more. But in a desolate part of the Sudan, say, or Burma, where extreme poverty is rife and famine just around the corner for many people, everybody is hungry and the fat people are probably the rich ones. Here variation in weight is caused by the environment, not the genes. In the argot of the scientist, the effect of the environment is non-linear: at the extremes, it has drastic effects. But in the moderate middle, a small change in the environment has a negligible effect.

The second surprise hidden in the average figures is that the influence of genes increases and the influence of shared environment gradually disappears with age. The older you grow, the less your family background predicts your IQ and the better your genes predict it. An orphan of brilliant parents adopted into a family of dullards might do poorly at school but by middle age could end up a brilliant professor of quantum mechanics. An orphan of dullard parents, reared in a family of double Nobel Prize-winners, might do well at school but by middle age may be working in a job that requires little reading or deep thought.

Numerically, the contribution of 'shared environment' to variation in IQ in a Western society is roughly 40 per cent in people younger than 20. It then falls rapidly to zero in older age groups. Conversely, the contribution of genes to explaining IQ variation rises from 20 per cent in infancy to 40 per cent in childhood to 60 per cent in adults and maybe even 80 per cent in people past middle age.[41] In other words, the effect of being reared in the same environment as somebody else is influential while you are still in that environment, but does not endure beyond the period of shared rearing. Adoptive siblings do have partly similar IQs while living together. But as adults their IQs are wholly uncorrelated. By adulthood, intelligence is like personality: mostly inherited, partly influenced by factors unique to the individual and very little affected by the family you grew up in. This is a counter-intuitive discovery exploding the old idea that genes come early and nurture late.

What this seems to reflect is that the intellectual experience of a child is generated by others. An adult, by contrast, generates his or her own intellectual challenges. The 'environment' is not some inflexible and real thing: it is a unique set of influences actively chosen by the actor himself. Having a certain set of genes predisposes a person to experience a certain environment. Having 'sporty' genes makes you want to practise at sport; having 'intellectual' genes makes you seek out intellectual activities. The genes are agents of nurture.[42]

As a parallel, how do genes affect weight? Presumably through controlling appetite. In an affluent society, those who gain most weight are hungrier and so eat more. The difference between a genetically fat and a genetically thin Westerner lies in the fact that the first is more likely to buy an ice cream. Is it the gene or the ice cream that causes his fatness? Well, it is obviously both. The genes are causing the individual to go out and expose himself to an environmental factor, in this case ice cream. Surely it is bound to be the same in the case of intelligence. The genes are likely to be affecting appetite more than aptitude. They do not make you intelligent; they make you more likely to enjoy learning. Because you enjoy it, you spend more time doing it and you grow more clever. Nature can only act via nurture. It can only act by nudging people to seek out the environmental influences that will fulfil their appetites. The environment acts as a multiplier of small genetic differences, pushing sporty children towards the sports that reward them, and pushing bright children towards the books that reward them.[43]

The main conclusion in behaviour genetics is counter-intuitive in the extreme. It tells you that nature plays a role in determining personality, intelligence and health – that genes matter. But it does not tell you that this role is at the expense of nurture. If anything, it proves rather dramatically that nurture matters just as much, though it is inevitably less good at discerning how (there is no environmental equivalent to the natural experiment created by identical and fraternal twins). Galton was utterly wrong in one important respect. Nature does not prevail over nurture; they do not compete; they are not rivals; it is not nature versus nurture at all.

Paradoxically, if Western society has reached the point where the heritability of intelligence is so high, then it means we have achieved something approaching a meritocracy, where your background does not matter. But this also reveals something truly surprising about genes. They do vary within the normal range of human behaviour. You might expect that genes would be like vitamin C or families – they only become limiting when they are malfunctional. So broken genes might cause rare broken minds, just as they cause rare diseases. Severe depression, mental illness or mental disability might be caused by rare variations in genes, just as all these things could be caused by rare and bizarre upbringings. This would then be the perfect utopia in which, so long as you had normal genes, and a normal family, everybody had the same potential personality and intelligence. The details would then be down to accident or circumstance.

But it is not like that. Behaviour genetics reveals very starkly that there are genetic differences that are common and that affect our personalities within the range of normal human experience. There are val–vals and met–mets among us, not just for the BDNF gene but for many other genes affecting personality, intelligence and other aspects of the mind. Just as some people are genetically better at gaining muscle strength than others, according to which version of the ACE gene they possess on chromosome 17,[44] so some people are genetically more able to absorb education according to which versions they possess of some unknown genes. These mutations are not rare; they are common.

From the point of view of the evolutionary biologist this is a scandal. Why is there so much 'normal' genetic variation, or to give it its proper name, polymorphism? Surely, the 'clever' variants on genes would gradually drive the 'dull' ones to extinction, and the phlegmatic ones would drive out the excitable ones. One kind must inevitably be superior to the other in providing survival or mating advantages. One kind must therefore endow its owner with greater ability to become a fecund ancestor. Yet there is no evidence of genes going extinct in this way. There seems to be a sort of happy coexistence of different versions of genes within the human population.

Enigmatically, there is more genetic variation in the human population than science has a right to expect. Behaviour genetics, remember, does not discover what determines behaviour; it discovers what varies. And the answer is that it is genes that vary. Contrary to popular opinion, most scientists love enigmas. They are in the business of finding new mysteries, not cataloguing facts. The white-coated ones in the labs live their lives in the dim hope of finding a really fine conundrum or paradox. And here is a fine one.

There are plenty of theories to explain the enigma, though none that is entirely satisfactory. Perhaps we human beings have simply relaxed natural selection so much by keeping ourselves alive with technology that our mutations have proliferated. But then why is the same variation present in other animals? Perhaps there is a delicate form of balancing selection that always favours the rare variants, thus keeping rare genes from going extinct. This idea certainly seems to explain variability in the immune system because disease favours rare versions of genes by attacking the common ones, but it is not immediately obvious why it should preserve a polymorphism in personality.[45] Perhaps mate choice encourages diversity. Or perhaps some new idea, as yet unheard of, will explain the phenomenon. Rival explanations for polymorphism were already causing bitter divisions among evolutionists in the 1930s and they are not settled yet.

ACCENTUATING THE POSITIVE

Normally at this point, a book about behaviour genetics would lurch into vitriolic criticism of one side or other of the nature–nurture argument. Either I would argue that twin studies are dubious in motive, flawed in design, idiotic in interpretation and likely to encourage fascism and fatalism. Or I would argue that they are a moderate and sensible corrective to the crazy dogma of the blank slate, which has forced us to try to believe that there is no such thing as innate personality or mental talent and that everything is the fault of society.

I have some sympathy with both views. But I am resolutely resisting the temptation to go in for this kind of comment, which has bedevilled the nature–nurture debate. The philosopher Janet Radcliffe Richards catches the gist nicely: 'If you follow up in detail any of the claims about what opponents are supposed to have said in this debate, you may be quite startled by the extent of misquoting, quoting out of context, looking for the worst interpretation of what is said, and flagrant misrepresentation that goes on.'[46] In my experience, scientists are most often wrong when they are being critical of each other. When they assert that their preferred idea is true and another idea is therefore false, they can be right in the first and wrong in the second: both ideas can be partly true. Like explorers arguing over which tributary is the source of the Nile, they are missing the point that the Nile needs both tributaries or it would be a creek. Any geneticist who says he has found an influence for genes, therefore there is no role for the environment, is talking bunk. And any nurturist who says he has found an environmental factor, therefore there is no role for genes, is talking equal bunk.

The IQ story contains a very clear example of this phenomenon. Called the Flynn effect, after its discoverer James Flynn, it is the remarkable fact that average IQ scores are rising steadily at the rate of at least five points per decade. This shows that the environment does influence IQ; it implies that compared with our grandparents we are all teetering on the brink of genius, which seems unlikely. None the less, something about modern life, whether it is nutrition, education, or mental stimulation, is making each generation better at IQ tests than its parents. Therefore, said one or two nurturists (not Flynn) triumphantly, the role of genes must be smaller than thought. But the analogy of height shows what a non sequitur this is. Thanks to better nutrition, each generation is taller than its parents, but nobody would argue that therefore height is less genetic than was thought. In fact, because more people now reach their full potential stature, the heritability of variation in height is probably increasing.

Flynn himself now thinks he understands his own effect by referring to the way appetite reinforces aptitude. During the twentieth century society progressively made it more rewarding for children to seek intellectual, school-based achievement. Thus rewarded, they responded by exercising those parts of the brain more. By analogy, the invention of basketball has encouraged more children to practise their basketball skills. As a result each generation is better at basketball. Two identical twins resemble each other in their basketball ability because they started out with a similar aptitude, which gave them the same appetite for the game, which brought them the same opportunities for practice. It is aptitude and appetite, not one or the other. An identical twin, having the same genes as his co-twin, therefore goes out and gets himself the same experience.[47]

EUTOPIA

Towards the end of his long life, Francis Galton succumbed to a temptation that befalls many prominent men. He wrote a utopia. Like all descriptions of the ideal society, from Plato's and Thomas More's onwards, it depicts the sort of totalitarian state that nobody in their right mind would want to inhabit. It is a useful reminder of a theme that recurs throughout this book: pluralism in the causes of human nature is vital. Galton was right about the strength of heritable factors in human nature, but wrong to think that nurture therefore does not matter.

Galton wrote his book in 1910 when he was in his eighties. It was called *Kantsaywhere* and it purports to be the diary of a man named Donoghue, a professor of vital statistics. He arrives in Kantsaywhere, a colony governed by a council along entirely eugenic lines. He meets Miss Augusta Allfancy, who is about to take an honours examination at the Eugenics College.

Kantsaywhere's eugenics policies were invented by a Mr Neverwas, who left his money to be used for the improvement of the human stock. Those who do well in the eugenic exams by having heritable gifts are rewarded in various ways; those who merely pass are allowed to breed in only a small way; those who fail are sent to labour colonies, where their duties are not especially onerous, but they must remain celibate. Propagation by the unfit is a crime against the state. Donoghue accompanies Augusta to various parties where she meets potential mates, for she will marry at 22.

Fortunately for Galton, Methuen rejected the novel for publication and his great-niece Eva managed to suppress it from wide circulation.[48] She at least realised how embarrassing it was. She could never have realised that Galton's controlled society would also be horribly prophetic for the twentieth century.

CHAPTER FOUR

The madness of causes

The word 'cause' is an altar to an unknown god.

William James[1]

During most of the twentieth century 'determinism' was a term of abuse, and genetic determinism was the worst kind. Genes were portrayed as implacable dragons of fate, whose plots against the damsel of free will were foiled only by the noble knight of nurture. This view reached its zenith in the 1950s, in the aftermath of the Nazi atrocities, but in some corners of philosophical inquiry it took hold much earlier. In psychiatry the fashion was swinging against biological explanations around 1900 at exactly the time that Galton was winning the argument for inheritance in human behaviour more generally. In view of what happened later, it is a bitter irony that this swing to nurture was happening first in the German-speaking world.

The central figure in the early history of psychiatry before Sigmund Freud was Emil Kraepelin. Born in 1856, Kraepelin trained as a psychiatrist in Munich in the late 1870s, but he did not enjoy the experience. He had bad eyesight and he disliked peering at slices of dead brain under microscopes. At the time psychiatry, a German speciality, was founded on the notion that the causes of mental illness would be discovered in the brain. If mind was the product of brain, then it followed that disorders of the mind could be traced to malfunctions of parts of the brain just as heart disease was caused by faulty parts of the heart. Psychiatrists were to become like heart surgeons, diagnosing and curing physical faults.

Kraepelin turned such reasoning on its head. After a period of academic migration, in 1890 he settled in Heidelberg and pioneered a new means of classifying mental patients based not on their current symptoms, let alone the appearance of their brains, but on their personal histories. He collected records on separate cards for separate patients, so that he could see the individual's history. Different mental illnesses, he argued, had characteristically different progressions. It was only by collecting information on each patient over a long period of time that you could begin to distinguish the separate features of each disease. Diagnosis was the child, not the father, of prognosis.

At the time, psychiatrists were seeing an increasing number of patients with a particular affliction. They were young, mostly in their twenties, and they suffered from delusions, hallucinations, emotional indifference and social insensitivity. Kraepelin was the first to delineate this apparently new illness, calling it *dementia praecox*, or precocious madness. It is now known by the even less helpful name coined in 1908 by Kraepelin's follower Eugen Bleuler – 'schizophrenia'. There is much argument today about whether schizophrenia was indeed suddenly more frequent or was just being noticed as mentally ill people for the first time emerged from the family and entered institutions. The balance of evidence suggests that despite such bias, there was a real increase in mental illness during the course of the nineteenth century and that schizophrenia in particular was a rare disease before the middle of the century.

Schizophrenia takes many forms, and varies in severity, but there are none the less remarkably consistent themes to the disease. Schizophrenics experience their thoughts as loud. In the old days, this was called hearing voices, but today it usually takes the form of believing that the CIA has implanted a device inside their heads. They also imagine that others can read their minds, and are apt to personalise every event, so that they think the television newsreader is sending them secret messages. Their paranoia leads to baroque conspiracy theories that render them likely to refuse treatment. Given how many ways the brain can go wrong, such a consistent pattern suggests that schizophrenia is a single disease, not a collection of similar symptoms.

Kraepelin distinguished *dementia praecox* from a different syndrome characterised by mood swings between mania and depression, which he called manic depression, and is nowadays called bipolar disorder. What was characteristic about each illness was its course and outcome, not its current manifestation. Still less could they be distinguished by visible differences in the brain. Kraepelin was saying that psychiatry should abandon anatomy and be agnostic about causes.

> As long as we are unable clinically to group illnesses on the basis of cause, and to separate dissimilar causes, our views about etiology will necessarily remain unclear and contradictory.[2]

But what is a cause anyway? The causes of human experience include genes, accidents, infections, birth order, teachers, parents, circumstance, opportunity and chance, to name just the most obvious. Sometimes one cause looms large, but not always. When you catch a cold the chief cause is a virus, but when you catch pneumonia the bacterium is only an opportunist – your immune system usually needs to be run down first by starvation, hypothermia or stress. Is that the 'true' cause? Likewise, 'genetic' diseases such as Huntington's chorea are caused precisely and simply by a mutation in one gene; environmental factors have almost no influence on the outcome. But phenylketonuria (PKU), a form of mental retardation caused by an inability to digest phenylalanine, could be said to be caused by the mutation, or by phenylalanine in the diet – it can be seen as either nature or nurture depending on your bias. How much more complex is the pattern when many different genes and many different environmental factors are almost certainly involved, as is probably the case with schizophrenia.

Therefore, in this chapter, by investigating the cause of schizophrenia, I hope to throw the whole notion of 'cause' into confusion. This is partly because the cause of schizophrenia is still very much an open question, with many rival candidate explanations covering all possibilities. You can still plausibly believe that genes, or viruses, or diets, or accidents are the first cause of psychosis. But the confusion goes deeper than that, for the closer science gets to understanding schizophrenia – and it is very close – the more it is blurring the distinction between cause and symptom. Environmental and genetic influences seem to work together, to require each other, till it is impossible to say which is cause and which is effect. The dichotomy of nature and nurture must first confront the dichotomy of cause and effect.

BLAME MOTHER

The first witness I call to explain the cause of schizophrenia is the psychoanalyst. For much of the middle part of the twentieth century he dominated the subject. Kraepelin's agnosticism about the causes of psychosis, which transfixed psychiatry at the turn of the twentieth century, left a vacuum which the Freudians were destined to fill. By apparently dismissing biological explanations of mental illness, and stressing life history, Kraepelin had opened the way for psychoanalysis, with its emphasis on childhood events as a cause of later neurosis and psychosis.

The extraordinary spread of psychoanalysis between 1920 and 1970 owed more to marketing than to therapeutic triumphs. By talking to patients about their childhood, analysts offered humanity and sympathy that had not been available before. This made them popular when the alternatives were a deep barbiturate sleep, insulin coma, lobotomy or electroshock convulsions: all unpleasant, addictive or dangerous. By emphasising the unconscious and the repression of memories from childhood, psychoanalysts also gave psychiatry 'a ticket out of the asylum'. Indeed, it could now offer its services to those who were not so much ill as unhappy, and who would pay well for the chance to recount their life story while lying on a couch. In the United States, thriving and lucrative private practice was the driving force by which psychoanalysts gradually took over the profession of psychiatry and made it their own. By the 1950s, even the training of psychiatrists was dominated by psychoanalysis. The key to each individual's psychological problems lay in his own individual history, and specifically in a social, or 'psychogenic' cause.

The 'talking treatment' was a great improvement on the contemporary alternatives. But, as is so often the case, psychoanalysis went too far and began to claim that other explanations were not only unnecessary, but wrong – morally as well as factually. Biological explanations of mental illness became heresy. Like all effective religions, psychoanalysis ingeniously redefined scepticism as further evidence of the need for its services. If a doctor prescribed a sedative or cast doubt on a psychoanalytic story, he was merely expressing his own neurosis.

At first Freudians avoided severe psychosis, concentrating instead on neurosis. Sigmund Freud himself was wary of treating psychotic patients, believing them to be beyond his methods, though he did hazard a wild guess that paranoid schizophrenia was the result of suppressed homosexual impulses. But as the confidence and power of analysts grew, especially in the United States, the temptation to tackle psychosis was irresistible. In 1935, a refugee analyst from Germany, Frieda Fromm-Reichmann, arrived at Chestnut Lodge in Rockville, Maryland, an institution already devoted to Freudian treatment. She quickly developed a new theory of schizophrenia: that it was caused by the patient's mother. In 1948 she wrote:

> The schizophrenic is painfully distrustful and resentful of other people due to the severe early warp and rejection he encountered in important people of his infancy and childhood, as a rule, mainly in a schizophrenogenic mother.[3]

Soon after this, a self-styled heir to Freud, Bruno Bettelheim, shot to fame with a similar diagnosis for autism: that it was caused by an indifferent 'refrigerator mother', whose coldness towards her son destroyed the boy's ability to acquire social skills. Bettelheim had been incarcerated by the Nazis in Dachau and Buchenwald, but he managed to bribe his way out of the worst parts of the camp and somehow arranged his own release in 1939, in circumstances that remain mysterious. He emigrated to Chicago, where he founded a home for emotionally disturbed children.[4] His enormous reputation did not long survive his suicide in 1990. Twin studies have utterly demolished the refrigerator mother theory, which spread guilt and shame among a generation of parents: the heritability of autism is 90 per cent. An identical twin with autism has an autistic co-twin in 65 per cent of cases; the concordance for fraternal twins is 0 per cent.[5]

Then it was the turn of homosexuals, this time the blame falling on the emotional stiffness of the father or the dominating personality of the mother. Some Freudians still cling to such theories. A recent book asserted:

> The father [of a gay man] is rejecting or withdrawn or weak or absent – emotionally, literally, or a combination of these – and the marital relationship is disharmonious. Gay men tend to have had negative relationships with their fathers, half of them (compared with a quarter of heterosexuals) feeling anger, resentment and fear towards fathers whom they deem cold, hostile, detached or submissive.[6]

All of which is probably true. It would be a miracle if most straight fathers did not have a 'negative relationship' with gay sons. But which came first? All but the most extreme Freudians have long since stopped assuming that the relationship causes the homosexuality, rather than vice versa. (The correlation tells you nothing about causality, let alone its direction.) The same is true of the parental theories of schizophrenia and autism. Mothers of autistic children, like fathers of homosexual boys, withdraw in frustration at the child's behaviour. Mothers of schizotypal children may indeed react badly to the child's developing psychosis. Consequence had been confused with cause.[7]

For the parents of schizophrenic youths, already under terrible stress, Freudian culpability was a blow they could have done without. The pain it was to cause to a generation of parents would have been more bearable if there was any evidence to support it. But it was soon obvious to any neutral observer that Freudian treatment was failing to cure schizophrenia. Indeed, by the 1970s some psychiatrists were brave enough to admit that psychoanalysis actually seemed to make the symptoms worse: 'The outcome for patients who received only psychotherapy was significantly worse than the outcome in the no-treatment control group,' said one, bleakly.[8] By then psychoanalysis had been used to treat tens of thousands of schizophrenics.

As so often in the middle years of the century the 'evidence' was based on a massive assumption – that nurture, not nature, explained most of the resemblance between parent and child. In the case of schizophrenia, had analysts not ignored biologists, they would have known that such an assumption was already unwarranted – because of studies of twins.

In the 1920s and 1930s a Jewish immigrant from Russia, Aaron Rosanoff, collected data on twins in California and used them to test the heritability of mental illness. Out of more than a thousand twin pairs in which one twin had mental illness, he identified 142 schizophrenics. In 68 per cent of the identical twins, the other twin also developed schizophrenia, whereas this was true of only 15 per cent of the fraternal twins. He found a similar difference in manic-depressive twins. Yet because genes were unfashionable in psychiatry, Rosanoff was ignored. According to the historian Edward Shorter:

> Rosanoff's twin studies arguably represent the major American contribution to international psychiatric literature in the years between the two world wars, yet the official histories of American psychiatry, dominated by psychoanalytically oriented writers, pass over his work in virtual silence.[9]

Franz Kallmann, who had emigrated from Germany in 1935, did a similar study of 691 twin schizophrenics in New York, and got an even stronger result (86 per cent concordance for identicals, 15 per cent for fraternals). He was howled down by the analysts at the World

Congress of Psychiatry in 1950. Rosanoff and Kallmann, both Jewish, were even accused of Nazism for using twin studies at all. The maternal theory of schizophrenia was shielded from uncomfortable facts for two more decades.

The current consensus is that 'psychosocial factors' are of tiny effect if they have any effect at all. In one Finnish study of adoptees, it was evident that the offspring of schizophrenics were slightly more likely to show thought disorder if their adoptive mothers were also showing what was euphemistically called 'communication deviance'. But there was no such effect for the offspring of unaffected biological parents. So if there is a 'schizophrenogenic mother' she can only affect those of her offspring with genetic susceptibility.[10]

BLAME THE GENES

The second witness to be called believes that schizophrenia is caused by genes. He uses all the arguments of behaviour genetics. Schizophrenia plainly runs in families. Having a first cousin with schizophrenia doubles your own risk from 1 per cent to 2 per cent. Having a half-brother or aunt with schizophrenia trebles it again to 6 per cent. Having a full sibling with the disorder puts you at 9 per cent risk. Having a non-identical twin raises the risk to 16 per cent. Having two parents with the disorder puts you at 40 per cent risk. And having a schizophrenic identical twin is the highest risk factor known for the disease: you then have roughly 50 per cent probability of also being schizophrenic yourself. (This number is considerably lower than in Rosanoff's and Kallmann's studies, because of more cautious diagnosis.)

But twins share nurture as well as nature. Beginning in the 1960s, Seymour Kety gradually demolished this objection with a growing study of Danish adoptees (Denmark has an unrivalled state database on children put up for adoption). He found that schizophrenia was ten times as common in the biological relatives of diagnosed schizophrenics who had been adopted as children as it was in their adopting families. The reverse experiment – children adopted by schizophrenics – is, of course, very rare.[11]

All these figures reveal two important things. First, that the heritability of schizophrenia in Western society is high: roughly 80 per cent, or about the same heritability as body weight and considerably more than personality. But second, they reveal that many genes are involved. Otherwise the fraternal twin figure would be much closer to the identical twin figure.[12]

The witness for genes is therefore remarkably convincing. Few diseases show such clear evidence of inheritance, except those that are caused by single genes. It ought to be a trivial matter, in this era of the genome, to identify the genes of schizophrenia. In the 1980s, full of confidence, geneticists set out to discover them. Schizophrenia genes were among the most popular quarry in the world of gene hunting. By comparing the chromosomes of people who have the disease with those of their relatives who do not, geneticists sought to pin down those bits of the chromosomes that were consistently different and so get a rough idea of where to look for the actual genes. By 1988, using the well-recorded pedigrees of Icelandic people, one team had a strong result. They had found a piece of chromosome 5 that was apparently abnormal in the schizophrenics, but not in their close relatives. About the same time a rival team stumbled on a similar phenomenon: schizophrenia apparently associated with having an extra piece of chromosome 5.[13]

Congratulations rained upon the winners. Headlines proclaimed the news of the finding of the 'schizophrenia gene'. It was one of many behaviour genes announced about this time – genes for depression, alcoholism and other psychiatric problems. The scientists themselves were careful to acknowledge in the small print that the result was preliminary, and that this was only one gene for schizophrenia, not *the* gene.

All the same, few were prepared for the disappointment that followed. Others tried to replicate the result without success. By the late 1990s, it was acknowledged that the association with chromosome 5 was a 'false positive' – a mirage. This has been the pattern with the genes affecting complex diseases of the mind: again and again over the past decade, they have proved illusory. Again and again, the initial excitement has faded. Scientists have learned to announce associations between a disorder and a chunk of one chromosome with much greater caution. Nobody now takes such an announcement seriously until it has been replicated.

Schizophrenia has now been linked to markers on most of the human chromosomes. Only six human chromosomes (3, 7, 12, 17, 19 and 21) do not have putative links to schizophrenia. But few of the links prove durable, and every study seems to find a different link. There could be good reasons for this. It could be that different populations have different mutations. The more genes that are involved in predisposing people to schizophrenia, the more likely it is that there will be different mutations producing similar effects. Imagine, for example, if the light goes out in your bedroom. It could be a failure of the light bulb, the fuse in the plug or the trip switch in the circuit; it might even be a power cut. Last time it was the trip switch; this time it proves to be the bulb. Failing to replicate an association between the trip switch and the fault, you indignantly reject it as a 'false positive'. Bulbs, not trip switches, are the cause of bedroom darkness.

Yet it could easily be both. In the brain, a system of far greater complexity, there are not three or four possible things that can go wrong, but thousands. Genes switch other genes on, which switch yet more genes on, and so on till there are scores of genes involved in even the simplest pathway. Knocking out any one could disrupt the whole pathway. But you would not expect the same gene knocked out in every schizophrenic. The more genes can cause the pathway to fail, the harder it will be to replicate associations between disease and gene. So false positives are not necessarily discouraging or even wrong (though some may be statistical flukes). Nor is the failure of linkage studies proof, as some have averred, that the whole concept behind 'neurogenetic determinism' is wrong. The role of genes in schizophrenia is proven by the twin and adoption studies, not by finding or failing to find particular genes. But it is fair to say that linkage studies, which worked so well for the single-gene diseases like Huntington's, have largely failed for psychoses.

BLAME THE SYNAPSES

Call the third witness. Some scientists, instead of trying to find what was different about the genes of schizophrenics, had set out to understand what was different about their brain biochemistry. From this they would then deduce which genes control this biochemistry and so investigate the 'candidate genes'. The first port of call was the dopamine receptor, dopamine being a 'neurotransmitter', or chemical relay system between certain neurons in the brain. One neuron releases dopamine into the synapse between cells (a synapse is a special narrow gap), and this causes the neighbouring neuron to begin transmitting electrical signals.

The focus on dopamine was inevitable after 1955, the year in which the drug chlorpromazine was first widely used on schizophrenics. To psychiatrists forced to choose between the brutality of a lobotomy and the uselessness of psychoanalysis, the drug was a godsend. It genuinely restored sanity. For the first time schizophrenics could leave the asylum and return to lead normal lives. Only later would the awful side effects of the drug emerge, and with them the problem of patients refusing to take their medication. Chlorpromazine induced in some patients a progressive degeneration of the control of movement similar to Parkinson's disease.

But if the drug was not a cure, it seemed to offer a vital clue to the cause. Chlorpromazine and its successors were chemicals that blocked dopamine receptors and prevented them having access to dopamine. Moreover, drugs that increase dopamine levels in the brain, such as amphetamines, provoke or exacerbate psychotic breaks. Third, brain imaging shows that the dopamine-fuelled parts of the brain are most atypical in schizophrenics. Schizophrenia must be a disorder of neurotransmitters, and in particular dopamine.

There are five different kinds of dopamine receptors on the receiving neurons. Two of these (D2 and D3) have proved to be faulty in some schizophrenics, but again the result is disappointingly weak and hard to replicate. Moreover, the best anti-psychotic drug prefers to block D4 receptors. To make matters worse, the D3 gene is on chromosome 3, which is one of the six chromosomes that have never managed to associate with schizophrenia in linkage studies.

The dopamine theory of schizophrenia gradually fell from fashion, not least after the discovery of mice with faulty dopamine signalling, which do not behave at all like schizophrenic people. Attention has recently focused on a different signalling system in the brain, the glutamate system. Schizophrenics seem to have too little activity at one kind of glutamate receptor (called the NMDA receptor) in their brains, just as they have too much dopamine. A third possibility is the serotonin signalling system. Here there has been better success, for one of the candidate genes, called 5HT2A, does seem to be quite often faulty in schizophrenics, and does sit on one of the chromosomes (13) most accused by linkage studies. But the effect is still disappointingly weak.[14]

Come the year 2000 neither linkage studies nor candidate gene searches had cracked the problem of which genes account for the heritability of schizophrenia. By then the Human Genome Project was nearing completion, so all the genes were at least present, laid out in the innards of computers, but how to find the few that matter? Pat Levitt and his colleagues in Pittsburgh sampled the prefrontal cortex of dead schizophrenics to find out which genes had been acting oddly. They carefully matched their subjects for sex, time since death, age and brain acidity. Then they used micro-arrays to sample nearly 8,000 genes and identify the ones that seemed differently expressed in schizophrenics. The first was a group of genes involved in 'presynaptic secretory functions'. In plain English this means the genes involved in producing chemical signals from neurons – signals like dopamine and glutamate. Two of these genes in particular were less active in the schizophrenics. Astonishingly, these genes are on chromosomes 3 and 17 – two of the six chromosomes where linkage studies had *not* found a schizophrenia association.[15]

But another gene also emerged from this study, which does map closely on to one of the right chromosomal spots (on chromosome 1). It is a gene called RGS4 and it is active on the downstream side of the synapse – that is, on the receiving end of the chemical signals. It was dramatically reduced in activity in the ten schizophrenics Levitt's group studied. In animals, RGS4 activity is reduced by acute stress. Perhaps this explains a universal feature of schizophrenics, that stress tends to bring on their psychotic episodes. In the case of the brilliant Princeton mathematician John Nash, an arrest and consequent loss of his job plus despair at failing to crack a problem in quantum mechanics seems to have tipped him over the edge. In Hamlet's case, seeing his mother marry his father's murderer might be thought enough stress to drive anybody mad. If such stress depresses the activity of RGS4, and if RGS4 is already low in vulnerable types, then stress could trigger the psychosis itself. But that would not mean that RGS4 is a cause of schizophrenia, just that its failure is a cause of worse symptoms in schizophrenics following stress – it is more like a symptom.

But curb even this much speculation with caution. The micro-array technique is picking up genes that have changed their expression in reaction to the disease, as well as genes that induce the disease. It could be confusing consequence with cause. Degrees of gene expression are not necessarily inherited. This is a vital issue that recurs throughout the book. Genes do not just write the script; they also play the parts.

However, the evidence from micro-arrays does at least support the hints from drug treatments that schizophrenia is a disease of the synapse, though it does little to sort cause from effect. Something is going wrong at the junctions between neurons in parts of the brain, most especially the prefrontal cortex.

BLAME THE VIRUS

Summon the fourth witness, who believes that schizophrenia is caused by a virus. The heritability of schizophrenia is high, he points out, but it is not total. Twin studies and adoption studies leave plenty of room for environmental factors to play a part. Indeed, they do more than that. They emphasise the role of nurture. No matter how many genes the geneticists eventually find, nothing will reduce the effect of the environment. Remember, nature is not at the expense of nurture; there is room for both, and they work together. Perhaps all that we inherit is a susceptibility, just as some people inherit a susceptibility to hay fever – but the cause of hay fever is surely pollen.

The twin studies reveal that an identical twin brother or sister of a schizophrenic has only a 50–50 chance of getting schizophrenia. Since the two have identical genes, there must be something non-genetic that halves the probability. Moreover, suppose the two identical twins have married different spouses and had children. As before, one twin then gets schizophrenia but the other does not. What will happen to the children? Clearly the children of the affected twin are at fairly high risk of schizophrenia, but what about the children of the twin who remains unaffected? You might expect that having escaped the disease himself, the unaffected twin is less likely to pass it on to his children. Yet this is not so. The children inherit the same risk from an unaffected parent, which proves that having the predisposing genes is necessary, but not sufficient, to develop the disorder.[16]

The search for the non-genetic factors in schizophrenia goes even farther back than the search for genes. It took a dramatic turn in 1988, however: the same year that the first genetic link was apparently found in Icelanders. This story, too, is Nordic, for while Robin Sherrington was testing the chromosomes of Reykjavik, Sarnoff Mednick was poring over medical records in the Helsinki Mental Hospital. Mednick was trying to explain a well-known fact about schizophrenia: more schizophrenics are born in winter than in summer. This is true in both hemispheres, despite the six-month difference in the date of the seasons. It is not a large effect, but it is undoubtedly there, and it refuses to go away, however the statistics are massaged.

Mednick's hunch was that influenza epidemics tend to occur in winter. Perhaps there is something about flu that predisposes mothers to give birth to potential schizophrenics. So he examined Helsinki hospital records to discover the effect of a 1957 influenza epidemic. Sure enough, those who had been in the middle three months of their own gestation during the epidemic were more likely to have the illness than those in the first or last trimester of gestation.

Mednick then read the obstetric records of women pregnant during the 1957 outbreak who gave birth to future schizophrenics. He found that they were more likely to have had the flu during the second trimester of pregnancy, the middle three months, than to have had it before or after. In Denmark meanwhile, a historical approach produced a supportive result: in those years between 1911 and 1950 when influenza had been rife, more schizophrenics had been born. And the riskiest date for the mother to catch the flu was in the sixth month, and especially the 23rd week, of her pregnancy.

So was born the viral hypothesis of schizophrenia: that influenza infection in pregnancy, especially during the second trimester, can cause some kind of damage to the immature brain which has the effect many years later of predisposing the affected person to psychosis. Of course, not all those whose mothers get influenza will become schizophrenics. The effect is bound to depend on genes: some people are genetically vulnerable to the impact of the virus, or infectiously vulnerable to the impact of their genes, whichever way you prefer to look at it.[17]

An intriguing hint that may support the influenza theory comes from the story of 'monochorionic' twins. About two-thirds of identical twins are even more intimately connected than the rest. They not only come from the same fertilised egg, but develop inside a single outer membrane or chorion within the womb and share the same placenta. (A few even develop within a single inner membrane, and are 'monoamniotic'.) The later the twinning event occurs, the more likely the twins are to be monochorionic. Since monochorionic twins are bathed in the same fluid during pregnancy, perhaps they encounter the same non-genetic influences. They even share blood through the common placenta. Perhaps they meet the same viruses. It would be especially interesting to know, therefore, if monochorionic twins are more concordant for schizophrenia than other identical twin pairs. Such data, however, are hard to gather. You would have to find not just twins, but schizophrenic twins whose birth records are available and sufficiently detailed to give an indication of whether they were in one bag or two. Not surprisingly, the data are just not available.

However, there are a few telltale signs. Some at least of the monochorionic twins show mirror-imaging: their hair swirls and fingerprints are on opposite sides, and they write with different hands. Further, the details of fingerprints are more similar in monochorionic twins: fingerprints are created in about the fourth month of gestation. Using these features as admittedly crude signs of monochorionic twins, James Davis in Missouri discovered a much higher concordance for schizophrenia in monochorionic twins than in dichorionic ones. He speculates that this may be evidence for the role of viruses, because the fluid-sharing twins are likely to share viruses as well. But the concordance of monochorionic twins might just indicate a shared exposure to accidental events of all kinds, not just infections.[18]

Other infectious agents, too, may be capable of triggering the chain of events that leads to susceptibility to schizophrenia, among them herpes virus and toxoplasmosis, a protozoan disease sometimes caught from cats. Toxoplasma can cross the placenta in a pregnant woman and blind or retard the foetus. It can also probably cause later schizophrenia. It has long been known that other insults to the developing foetus may be risk factors for schizophrenia, including especially birth complications. The facts are hard to interpret because schizophrenic mothers are prone to birth complications themselves. None the less it seems that a foetus starved of oxygen in the womb by pre-eclampsia is at nine times the normal risk of schizophrenia. What the medical fraternity delicately calls hypoxic insults – near-suffocation – during birth is a definite risk factor. Again it seems to interact with genes. You can endure a hypoxic episode better with the right genes, or you can outwit your genetic fate better with an easy birth.[19]

Hypoxia may be a reason for the fact that twins do not have identical risks, even though they share the predisposing genes. During birth, or before it, one twin may be more likely to experience hypoxia than the other. That may be why they do not both show the disease in later life.

However, there is another, more intriguing possibility. The virus that causes AIDS is a retrovirus, which means that when you catch AIDS, the genes of the virus are literally incorporated into the DNA in the chromosomes of some of your cells. Because this happens in blood cells and not in sperm or egg cells, such genes cannot be passed on to your offspring. But some time in the distant past – and more than once – a similar retrovirus has managed to infect germ cells. We know this because the human genome contains many different copies of complete retroviral genomes, recipes for making infectious viral particles. Known as hervs (for human endogenous retroviruses), they sit among our own genes, parasitic intruders, and we pass them on to our offspring. Indeed, simplified and abridged versions of these viral genomes are among the commonest motifs in our genome – they are the so-called jumping genes that make up nearly a quarter of our DNA. We human beings are, at the DNA level, substantially descended from viruses.

Luckily, the viral DNA is kept under a sort of house arrest, shut down by a mechanism called methylation. But there is always the risk of a herv escaping, making a virus and infecting our cells from within. If that were to happen, the medical effect would be bad enough, but consider what philosophical damage it would also do – to the nature–nurture debate. It is an infectious disease, just like any other virus, but it starts within our very own genes, and is passed on from parent to child as a set of genes. It would look like an inherited disease, but behave like an infection.

A few years ago, evidence began to emerge that precisely such an event might explain multiple sclerosis. MS is quite unlike schizophrenia in symptoms, but it shares a few coincidences. Both occur in early adulthood; both are more frequent in people who were born in winter. So Paromita Deb-Rinker, a Canadian scientist, analysed the DNA from three pairs of identical twins in which one member of the pair had schizophrenia and the other did not. By comparing the DNA from the affected twins with that from the unaffected twins, she found evidence of a herv that might be more active, or present in more copies, in the affected twin.[20] Robert Yolken and his colleagues at Johns Hopkins University also looked for evidence of herv activity in schizophrenics. They tested the cerebrospinal fluid from 35 people newly diagnosed with schizophrenia in Heidelberg in Germany, 20 people who had suffered from the disorder for many years in Ireland, and 30 healthy controls from the same two places. Ten of the German schizophrenics, one of the Irish ones and none of the controls had evidence of active herv genes. Moreover, the retrovirus that was active was from the same family of herv as the one associated with multiple sclerosis.[21]

None of this yet proves that hervs are relevant to the disease, let alone the cause, but they do suggest a connection. If hervs were indeed causing schizophrenia, perhaps themselves triggered by influenza infection in the womb, and perhaps by interfering with other genes during the development of the frontal cortex of the brain, then it would explain why the disorder is both highly heritable and apparently associated with different genes in different people.

BLAME DEVELOPMENT

The fifth witness brings a mouse. This is no ordinary mouse, but one that behaved rather oddly in its cage some time in 1951. It moved with a strange 'reeling' motion, as if dancing (but not in a way to be confused with the Japanese waltzing mice I mentioned in chapter 2). A scientist duly noticed the phenomenon, and by back-crossing quickly proved that the cause was a single gene inherited from both parents. The brain of the reeler mouse is a bit of a mess, principally because some layers of cells that should be on the inside are on the outside instead. The 'reelin' gene was located in 1995 on the mouse's fifth chromosome, and the human equivalent soon followed in 1997: a gene on chromosome 7 that produced a protein 94 per cent homologous with the mouse protein. It is a very big gene, with more than 12,000 letters, divided into no less than 65 separate 'paragraphs' called exons. Subsequent experiments have shown that reelin protein is vital to the organisation of the brain in the foetus of both a mouse and a human being. It directs the organised formation of layers in the brain, apparently by telling neurons where to grow to and when to stop.

What has all this to do with schizophrenia? In 1998 a team at the University of Illinois measured the quantity of reelin in the brains of recently dead schizophrenics and found that it was half that of normal dead brains.[22] A new potential suspect entered the frame. Disordered neuronal migration is a characteristic of schizophrenia, and reelin is one of the organisers of neuronal migration. It also helps to maintain the 'dendritic spines' at which synapses form, so a shortage could lead to faulty synapses. For devotees of the influenza theory, it quickly became apparent that one way to cause a transient 50 per cent reduction in reelin expression in the brain of a mouse was to give it a prenatal infection with human influenza.[23] In other words, reelin seemed to tie together the other theories of schizophrenia.[24]

The poor reeler mouse immediately became the focus of much attention: perhaps it would prove to be an animal model of schizophrenia. The reeling behaviour is only apparent if the mouse has inherited the faulty gene from both parents. If it has only one faulty gene, a mouse seems superficially normal. But it is not. It learns its way through a maze much more slowly and never gets as good at the task as a normal mouse. It is less sociable than normal mice.

This is hardly rodent schizophrenia, but perhaps it has a few parallels. Hopes that reelin would prove to be the chief cause of schizophrenia began to fade, however, in the 1990s when human reelers were discovered in two separate families in Saudi Arabia and England. In both these families cousins had married each other and the marriage had brought together faulty versions of the reelin gene, causing a disorder called lissencephaly with cerebellar hypoplasia (LCH), which is usually fatal within four years of birth. If inherited reelin deficiency is the cause of schizophrenia, then you would expect that some of the apparently unaffected relatives of these unfortunate children would be schizophrenic, because they are carrying the mutation in one of their genes. But so far there is no history of schizophrenia in either family, though the Arab family has not been studied in detail. Once again, as so often in the schizophrenia story, a promising start leads to a dead end. Reelin reduction is part of schizophrenia, perhaps a crucial part, but probably not one of the primary causes.[25]

Bizarrely, reelin reduction is not confined to schizophrenia, but is common in patients with severe bipolar depression and autism as well. It is almost as if a reduction in reelin can cause different brain problems depending on where in the brain, or when during development, it occurs. Reelin and influenza both point the finger at events in the womb, which is at first sight puzzling because the most characteristic feature of schizophrenia is that it is a disease of adults. Although children who will later become schizophrenics can be identified retrospectively as anxious, slow to walk and poor at verbal comprehension,[26] most are by no means ill until after puberty. How can a disease be caused in the womb and expressed in adulthood?

The neurodevelopmental model of schizophrenia attempts to explain this conundrum. In 1987 Daniel Weinberger argued that schizophrenia was unlike other brain disorders in that the cause was no longer there when the symptoms appeared. The damage had been done much earlier, but only became apparent because of some later, normal brain maturation process: the early effects are 'unmasked' by later development as adulthood approaches. Unlike, say, Alzheimer's or Huntington's disease, schizophrenia is not a disease of brain degeneration, but a disease of brain development.[27] For example, during late adolescence and early adulthood the brain is extensively altered. Many of its wires are insulated for the first time, and many of its connections are 'pruned': synapses between neurons are cut back, leaving only the strongest ones. Perhaps in schizophrenics either there is too much pruning in the prefrontal cortex in reaction to a failure of the synapses to develop properly many years before, or perhaps too few neurons have migrated or extended to their targets. There will be many genes that mitigate or exacerbate these effects, or maybe respond to them, and they might therefore be termed 'schizophrenia genes', but they are more like symptoms than causes. It is among the genes that affect the original early development that one must seek for true 'causes' of schizophrenia.[28] (It is perhaps no coincidence that schizophrenia appears at the age when young men and women are competing most fiercely to gain footholds in an unfamiliar adult world and win mates.)

Most scientists are agreed that in this sense schizophrenia is an organic disease, a disease of development, a disease of the fourth dimension, the dimension of time. It is caused by something going awry in the normal growth and differentiation of the brain. It is another forceful reminder that bodies – and brains – are not made, like model aeroplanes. They are grown, and that growth is directed by genes. But those genes react to each other, to environmental factors and to chance events. To say that genes are nature and the rest is nurture is almost certainly wrong. Genes are the means by which nurture expresses itself, just as surely as they are the means by which nature expresses itself.

BLAME THE DIET

But no lover of science should ever be happy when things are getting consensual. The sixth witness is determined to upset the mood. He believes that genes, development, viruses and neurotransmitters all play a part, but none is the really fundamental explanation of the cause. All are really symptoms. The key to understanding schizophrenia, he asserts, lies in what we eat. In particular, the developing human brain has a desperate need for certain fats, known as essential fatty acids, and the brains of schizotypal people need more of these than usual. If they do not get them in their diet, the result can be schizophrenia.

In February 1977, on a bright but bitterly cold day, a British medical researcher was walking through Montreal when he had his own 'eureka' moment. David Horrobin had been trying to fit together pieces of a mental jigsaw of odd facts about schizophrenia. They all related to the often-forgotten non-mental aspects of the disease, and they were these. First, schizophrenics rarely suffer from arthritis; second, they are surprisingly insensitive to pain; third, their psychosis often gets much better, temporarily, when they have fevers (astonishingly, malaria was once tried as a cure for schizophrenia – it worked, but only temporarily). The fourth piece of the jigsaw in Horrobin's head was new. He had just noticed that a chemical called niacin, then used to treat high cholesterol, did not cause a flushing of the skin in schizophrenics as it did in other people.[29]

Suddenly the pieces all fitted together. The skin flushing, the inflammation in arthritis and the pain response all depend on the release of fat molecules called arachidonic acid (AA) from the membranes of cells. These are converted into prostaglandins, which cause some of the signs of inflammation, redness and pain. Likewise, a fever also releases AA. So perhaps schizophrenics were unable to release normal quantities of AA from their cells and this caused their mental problems as well as their resistance to pain, arthritis and flushing. Only a dose of fever raised their AA levels to those seen in normal people and restored their normal brain function. Horrobin duly published his hypothesis in the *Lancet* and sat back to wait for the applause. There was a deafening silence. The schizophrenia experts were too immersed in the dopamine hypothesis at that time even to notice a different theory, let alone consider it. Schizophrenia was brain disease, so what was the relevance of all this stuff about fats?

Horrobin likes to defy conventional wisdom and he was undaunted. But it was not until the 1990s that evidence started to come in supporting his hunch. Deficits in AA in schizophrenics were soon reported, as was an increased rate of oxidation of AA. Details gradually emerged from the fog of ignorance suggesting that either AA leaks too easily from the cell membranes of schizophrenics, or AA once released cannot be incorporated back into membranes easily – perhaps both.

Both processes are the result of faulty enzymes, and enzymes are made by genes, so Horrobin is happy to allow a role for genes in predisposing people to schizophrenia. But in expressing the disease, or better still, curing it, he believes that diet may play a role.

A learned and lengthy disquisition on the nature and function of fats and fatty acids is probably necessary at this point. But I fear the reader did not buy this book because she was in love with biochemistry. So I am going to try to boil down the essential facts about fats into a few terse sentences. Each cell in your body is held together by an outer membrane, which is made largely of fat-rich molecules called phospholipids; a phospholipid is like a three-pronged fork, each prong being a long fatty acid. There are hundreds of different fatty acids to choose from, ranging from saturated to polyunsaturated, the key feature of the polyunsaturated ones being that they make a more flexible prong. This matters especially in the brain, where the membrane of a brain cell must not only adopt an intricate shape, but also change rapidly as connections between cells are added or lost. So the brain needs more polyunsaturated fatty acids than other tissues: about one-quarter of its dry weight consists of just four kinds of polyunsaturates. They are known as the essential fatty acids (EFAs) because our neglectful ancestors never invented the ability to make them from scratch; their precursors come from food, having worked their way up the food chain from the simple algae and bacteria that do know how to make them. If one person eats a diet rich in saturated fats and poor in EFAs, he may end up with brain cell membranes that are less flexible than those of somebody who eats a lot of fatty fish. (This does not easily explain how schizophrenia is just as common in countries like Norway and Japan, where fish forms a large part of the traditional diet.)

The obvious test of Horrobin's ideas is to treat schizophrenics with EFAs. His colleague Malcolm Peet and others have begun to do so. The results are not spectacular, but they are encouraging. A large daily dose of fish oil – rich in EFAs – does produce a modest improvement in the symptoms of schizophrenics. In 31 newly diagnosed Indian schizophrenics, a dose of one of the four main EFAs, called EPA, had such an effect in a double-blind trial (where neither the doctor nor the patient knew which patients were getting the drug until afterwards) that ten no longer needed to take anti-psychotic drugs to control their illness; none of the 29 control subjects given the placebo saw any improvement. EPA inhibits the enzyme that removes arachidonic acid from neuronal membranes; it therefore preserves the AA in the membrane. Since most anti-psychotic drugs have pretty dreadful side effects, from listlessness and weight gain to the symptoms of Parkinson's disease, this is exciting news.

The fatty acid story is not a rival to the various genetic hypotheses. Many of the neural symptoms of schizophrenia could be connected to fatty acids. EFAs are known to regulate the pruning of neuronal connections at puberty. Women are better at making EFAs from their dietary precursors, and women are less likely to get schizophrenia. Starvation during pregnancy, hypoxia during birth, stress and even influenza infection have all been shown to reduce the availability of EFAs to the developing brain. The flu virus actually inhibits the formation of AA, possibly because AA is needed as part of the body's defence.

More direct evidence for the fatty acid theory comes from some of the actual genes implicated in schizophrenia. They include the gene for phospholipase-A_2, a protein whose job is to remove the middle prong of the phospholipid fork, the one that is usually an EFA. The gene for apoD, a sort of delivery truck that brings fatty acids to the brain, is three times as active in schizophrenics in the very part of the brain most implicated in the symptoms of the disease – the prefrontal cortex – but not in the rest of the brain or body. It is almost as if the prefrontal cortex, finding itself short of these fatty acids, cranks up the expression of the apoD gene in an attempt to compensate (the apoD gene, by the way, is on chromosome 3, where no 'schizophrenia gene' was detected by the linkage studies). One of the reasons that clozapine is an effective drug against schizophrenia might be its ability to encourage the expression of apoD. Horrobin's hypothesis is that for full schizophrenia you require two genetic faults: one that reduces your ability to incorporate EFAs into cell membranes; and another that takes them out too easily (each could be affected by several genes). Even with both these genetic faults, an outside event is also required to trigger the psychosis, and other genes can modify or even forbid the effect.[30]

METHOD IN OUR MADNESS

Schizophrenia is about equally common all over the world and in all ethnic groups, occurring at the rate of about one case per hundred people. It displays much the same form in Australian Aborigines and Inuits.[31] This is unusual; many genetically influenced diseases are either peculiar to certain ethnic groups or much commoner in one group than another. It implies perhaps that the mutations that predispose some human beings to schizophrenia are ancient, having occurred before the ancestors of all non-Africans left Africa and fanned out across the world. Since being schizophrenic is hardly conducive to survival, let alone successful parenthood, in a Stone-Age world, this universality is puzzling: why have the genetic mutations not died out?

Many people have noticed that schizophrenics seem to appear in successful and intelligent families. (Such an argument led Henry Maudsley, the British contemporary of Kraepelin, to reject eugenics, because he realised that sterilising those with a taint of mental illness would wipe out a lot of geniuses, too.) People with a mild version of the disorder – sometimes called 'schizotypal' people – are often unusually brilliant, self-assured and focused. As Galton put it, 'I have been surprised at finding how often insanity has appeared among the near relatives of exceptionally able men.'[32]

This eccentricity may even help them to great success. It is perhaps no accident that many great scientists, leaders and religious prophets seem to walk the crater rim of the volcano of psychosis, and to have relatives with schizophrenia.[33] James Joyce, Albert Einstein, Carl Gustav Jung and Bertrand Russell all had close relatives with schizophrenia. Isaac Newton and Immanuel Kant might both be described as 'schizotypal'. One absurdly precise study estimates that 28 per cent of prominent scientists, 60 per cent of composers, 73 per cent of painters, 77 per cent of novelists, and an astonishing 87 per cent of poets have shown some degree of mental disturbance.[34] As John Nash, the Princeton mathematician, said after recovering from 30 years of schizophrenia and accepting a Nobel Prize for his work on game theory, the interludes of rationality between his psychotic episodes were not welcome at all. 'Rational thought imposes a limit on a person's concept of his relation to the cosmos.'[35]

The Michigan psychiatrist Randolph Nesse speculates that schizophrenia may be an example of an evolutionary 'cliff effect', in which the mutations in different genes are all beneficial, except when they all come together in one person, or evolve just too far, at which point they suddenly combine to produce a disaster. Gout is a cliff disease of this kind. High levels of uric acid in the joints protect human beings from premature ageing, but a few people get too much of it and painful crystals of the stuff form in their joints. Perhaps schizophrenia is the result of too much of a good thing: too many genetic and environmental factors that are usually good for brain function all coming together in one individual. This would explain why the genes predisposing people to schizophrenia do not die out; so long as they do not combine, they each benefit the survival of the carrier.

MENTAL CONFUSION

During the twentieth century the ideological forces of nature and nurture often behaved like medieval armies laying siege to diseases as if they were castles. Scurvy and pellagra, explained as vitamin deficiencies, fell to the forces of nurture, while haemophilia and Huntington's, explained as genetic mutations, fell to the army of nature. Schizophrenia was a vital border stronghold, held by nurture for much of the century as a fortress of Freudian theory. But although the Freudians – those Knights Templar of the nature–nurture war – were driven from the battlements decades ago, the geneticists have never managed to occupy the fortress convincingly and they may be forced to call a truce and welcome nurturist forces back over the moat.

A century after the syndrome was first identified, the only two things that can be said for certain about schizophrenia are that blaming unemotional mothers was plain wrong, and that there is something highly heritable about the syndrome. Beyond that, almost any combination of explanations is possible. Many genes clearly influence susceptibility to schizophrenia, many may respond to it in compensation, but few seem to cause it. Prenatal infection seems to be vital in many cases, but it may be neither necessary nor sufficient. Diet can exacerbate the symptoms and perhaps even trigger the onset of them, but probably only in those who are genetically susceptible.

In tackling psychosis, neither nature theories nor nurture theories are much good at distinguishing cause from effect. The human brain is wired to seek simple causes. It eschews uncaused events, preferring instead to deduce that when A and B are seen together, either A causes B or B causes A. This tendency is strongest in schizophrenics, who see causal connections between the most patent coincidences. But often A and B are simply parallel symptoms of something else. Or even worse, A can be both the cause and the effect of B.

Here then is a perfect illustration that nature and nurture both matter. I promised that schizophrenia would confuse the issue and it has. Kraepelin was wise to be agnostic about the cause: even with all the weight of modern science behind them his successors have failed to find it. They have even failed to distinguish cause from effect. Instead, it looks highly possible that the ultimate explanation of schizophrenia will include both nature and nurture, neither of which will be able to claim primacy.

CHAPTER FIVE

Genes in the fourth dimension

If we follow a particular recipe, word for word, in a cookery book, what finally emerges from the oven is a cake. We cannot now break the cake into its component crumbs and say: this crumb corresponds to the first word in the recipe; this crumb corresponds to the second word in the recipe.

Richard Dawkins[1]

The job of curator of the mollusc collection at the natural history museum of Geneva is not to be sniffed at. When it was offered to Jean Piaget, he was well qualified, having published nearly 20 papers on snails and their cousins. But he turned it down, and for a good reason: he was still a schoolboy. He went on to do a doctorate on Swiss molluscs before his godfather, alarmed at his obsession with natural history, diverted him from malacology to philosophy first in Zurich and then at the Sorbonne. But his fame rests on his third career, begun at the Rousseau Institute in Geneva in 1925: as a child psychologist. Between 1926 and 1932, still precocious, he published five influential books on the minds of children. It is to Piaget that modern parents owe their obsession that little Johnny must meet his developmental milestones.

Piaget was not the first person to observe children as if they were animals – Darwin did the same with his own children – but Piaget was probably the first to think of them not as apprentice adults, but as a species equipped with a characteristic mind. The 'errors' five-year-old children made in answer to questions in intelligence tests revealed to Piaget the peculiar but consistent ways in which their minds worked. In trying to answer the question 'how does knowledge grow?', he saw a progressive, cumulative construction of the mind during childhood in response to experience. Each child goes through a series of developmental stages, always in the same order, though not always at the same rate. First comes the sensorimotor stage, when the infant is little more than a bundle of reflexes and reactions; it cannot yet conceive that objects still exist when hidden. Next comes the pre-operational stage, an age of egocentric curiosity. Then the stage of concrete operations. And last, on the brink of adolescence, the dawn of abstract thought and deductive reasoning.

Piaget realised that development was more continuous than this implies. But he insisted that, just as a child will not walk or talk until he is 'ready', so the elements of what the world calls intelligence are not merely absorbed from the outside world; they appear when the developing brain is ready to learn them. Piaget saw cognitive development neither as learning, nor as maturation, but as a combination of the two, a sort of active engagement of the developing mind with the world. He thought the mental structures necessary for intellectual development are genetically determined, but the process by which the maturing brain develops requires feedback from experience and social interaction. That feedback takes two forms: assimilation and accommodation. A child assimilates predicted experiences and accommodates to unexpected ones.

In nature–nurture terms, Piaget, alone among the men in my photograph, defies categorisation as an empiricist or a nativist. Where his contemporaries Konrad Lorenz and B.F. Skinner took up extreme positions, the first as a champion of nature, the second of nurture, Piaget picked a careful path right through the middle. With his emphasis on development through stages, Piaget vaguely prefigured the ideas of formative experiences in youth. He was wrong in many particulars. His hypothesis that a child understands the spatial properties of objects only by handling them has been disproved. Spatial understanding seems to be much closer to innate than that – even very small babies can understand spatial properties of things they have never handled. None the less, Piaget deserves some credit for being the first to take seriously the fourth dimension of human nature – the time dimension.[2]

THE EXCESSES OF NATIVISM

It was this concept, rediscovered a little later by zoologists, that came to play a central role in one of the most illuminating of the nature–nurture debates, the one between Konrad Lorenz and Daniel Lehrman in the 1950s and 1960s. Lehrman was an ebullient and articulate New Yorker with a passion for birdwatching, who made a discovery about the behaviour of ring doves that had broad implications for human beings as well. He found that the male dove's courtship dance triggers a change in the female dove's hormones. Thus, an external experience can cause, via the nervous system, an internal, biological change in the organism. He did not know it, but such a response is mediated by the switching on and off of genes.

In 1953, before the climax of his dove work, Lehrman decided to use his halting German, learned while decoding radio intercepts for American intelligence in the Second World War, to translate Lorenz's work into English – in order to criticise it. His powerful critique was to influence a generation of ethologists. Even Niko Tinbergen would moderate his views after reading Lehrman. The Austrian Lorenz had been championing instinct – the idea that some behaviour is innate in the sense that it will emerge even if the animal is insulated from its normal environment from birth. Most animals, said Lorenz, were driven to elaborate and sophisticated behaviour patterns, not by their experience, but by their genes. In his critique Lehrman charged that Lorenz had omitted all mention of development: of how the behaviour came to be. It did not spring fully formed from the gene; the genes built a brain, which absorbed experience before it emitted behaviour. In such a system, what is meant by the word innate?[3]

Lorenz replied at length, and Lehrman responded again, but the two were largely at cross-purposes. Lehrman argued that just because a behaviour is the product of natural selection, that does not mean it is 'innate', meaning produced without experience. Before a dove can develop a mate preference for its own species, it needs to experience a parent dove; the same is not true in a cowbird, which like a cuckoo never sees its parents and therefore does have truly 'innate' mate preferences. Lorenz hardly cared how the behaviour was produced so long as it was obviously selected by natural selection and expressed in the adult animal in much the same way given normal experience. For him, innate meant inevitable. Lorenz was always going to be more interested in the why than the how.

Tinbergen resolved the issue to the satisfaction of many when he said that a student of animal behaviour should ask four different questions about a particular behaviour. What are the mechanisms that cause the behaviour? How does the behaviour come to develop in the individual (Lehrman's question)? How has the behaviour evolved? What is the function or survival value of the behaviour (Lorenz's question)?[4]

The argument was cut short by Lehrman's death in 1972. Yet, in recent decades Lehrman's developmental argument has become something of a standard to rally those who think the nativists of behaviour genetics and evolutionary psychology have gone too far. The 'developmentalist challenge' takes many forms, but its central charges are that many modern biologists talk much too glibly about 'genes for' behaviour, ignoring the uncertainty, complexity and circularity of the system through which genes come to influence behaviour. According to the philosopher Ken Schaffner, a five-point manifesto of the developmentalist challenge might go something like this: genes deserve parity with other causes; they are not 'preformationist'; their meaning depends heavily on context; the effects of genes and environments are seamless and inseparable; and the psyche 'emerges' unpredictably from the process of development.[5]

In its strongest form, as presented by the zoologist Mary Jane West-Eberhard, the challenge claims to present a 'second evolutionary synthesis' that will overthrow the first – the fusion of Mendel and Darwin that came about in the 1930s – by elevating the mechanisms of development alongside those of genetics.[6] For instance – and this is my example – take a glance at the pattern of blood vessels on the back of your hands. Although the veins get to the same destinations on both hands, they get there by slightly different routes. This is not because there are different genetic programmes for the different hands, but because the genetic programme is flexible: in some way it delegates local steering to the vessels themselves. Development accommodates to the environment: it is capable of coping with different circumstances and still achieving a result that works. If different developments can result from the same set of genes, then different genes might also be capable of achieving the same outcome. Or to put it in technical terms, development is well 'buffered' against minor genetic changes. This might explain two intriguing phenomena. First, wild breeds, such as wolves, are much less sensitive to individual genetic mutations than inbred forms such as pedigree dogs: they are buffered by their genetic variation. In turn, this might explain the otherwise puzzling fact that there are so many different versions of each gene about in the population (in human beings as well as other wild animals). Many genes come in two slightly different versions, one on each equivalent chromosome, which may help to provide the flexibility to develop a working body in different environments.

The development of behaviour need be no less flexible and buffered than the development of anatomy.[7] In its weaker form, the developmentalist challenge is merely a reminder to behaviour geneticists not to draw conclusions that are too simplistic, and not to encourage newspaper headline writers to speak of 'gay genes' or 'happiness genes'. Genes work in huge teams and build the organism and its instincts not directly but through a flexible process of development. Those who actually study genes and behaviour – in mice, flies and worms – say they are well aware of the dangers of oversimplification, and are sometimes a little irritated by the developmentalists. As much as they emphasise its complications and flexibility, even development is still at root a genetic process. Experiments confirm the complexity, plasticity and circularity of the system, but also reveal that even the environment affects development only by switching genes on and off: genes that allow plasticity and learning. Ralph Greenspan, a pioneer of the study of fruit fly courtship, put it this way:

Just as the ability to carry out courtship is directed by genes, so too is the ability to learn during the experience. Studies of this phenomenon lend further support to the likelihood that behavior is regulated by a myriad of interacting genes, each of which handles diverse responsibilities in the body.[8]

IN THE KITCHEN

Once you try to think about the fourth dimension of the organism, several useful parables come to mind, all of them rather graphic. Metaphor, in my view, is the lifeblood (ha!) of good scientific prose, so I shall explore two of these parables at length.

The first is the parable of canalisation, coined by the British embryologist Conrad Waddington in 1940.[9] Consider a ball at the top of a hill. As it rolls down, the hill is smooth at first, but after a while gullies begin to appear in the surface; before long the ball is rolling down a narrow channel. On some hills the gullies converge into one channel; on others, they diverge into several channels. The ball is the animal. The hill of the converging gullies represents the development of the most 'innate' kind of behaviour: it will always turn out roughly the same whatever the organism's experience. The hill of the diverging gullies represents behaviour that is much more 'environmentally' determined. Yet both still require genes, experience and development to appear at all. So, for instance, grammar is highly canalised; vocabulary is not. The formulaic song of a wren – which I just heard outside my window – is much more canalised than the imitative and inventive song of the thrush I can also hear.[10]

Equating innate behaviour with canalised development is a useful, if limited, idea, not least because it cuts across the gene–environment dichotomy so cleanly: something can be well specified by genes and still thrown into a different channel by the environment. If personality and IQ are highly heritable in most kinds of society (chapter 3), this implies that their development is narrowly canalised – it would take a very different environment to throw the ball so far off track that it ends up in a different channel. But it does not mean that the environment is unimportant: the ball still needs a hill to roll down.

For my next sermon, I will expatiate upon a different parable, one that dates from 1976, when it was coined by Pat Bateson, a British ethologist much influenced by Lehrman. This is the parable of the kitchen:

> The processes involved in behavioural and psychological development have certain metaphorical similarities to cooking. Both the raw ingredients and the manner in which they are combined are important. Timing also matters. In the cooking analogy, the raw ingredients represent the many genetic and environmental influences, while cooking represents the biological and psychological processes of development.[11]

The kitchen analogy has proved popular with both sides of the nature–nurture argument. Richard Dawkins used the metaphor of baking a cake in 1981, while emphasising the role of genes; his arch-critic Steven Rose used the very same metaphor three years later while arguing that behaviour is 'not in our genes'.[12] The cooking metaphor is not perfect – it fails to capture the alchemy of development in which two ingredients lead automatically to the production of a third and so on – but it deserves its popularity, for it expresses so well the fourth dimension of development. As Piaget noticed, the development of a certain human behaviour takes a certain time and occurs in a certain order, just as the cooking of a perfect soufflé requires not just the right ingredients, but also the right amount of cooking and the right order of events.

Likewise, the cooking metaphor instantly explains how a few genes can create a complex organism. Douglas Adams, the science-fiction writer, sent me an email shortly before his untimely death, criticising the argument that 30,000 genes were too few to specify human nature. He suggested that the blueprint of a cake, such as an architect would need, would indeed be an immensely complicated document, requiring an exact vector for each raisin, an exact description of the shape and size of each dollop of jam, and so on. If the human genome were a blueprint, then even 30,000 genes would never be sufficient to specify a body, let alone a psyche. The recipe for a cake, on the other hand, is a simple paragraph. If the genome were a recipe – a set of instructions for 'cooking' the raw ingredients in certain ways for certain lengths of time – then 30,000 genes would be ample. One cannot only imagine such a process in the growing of a limb, one can now actually see the rudiments of how it works, gene by gene, emerging from the scientific literature.

But can you imagine such a thing for behaviour? Most people's minds so boggle at the thought of molecules, made by genes, generating an instinct in the mind of a child that they give up and call the process impenetrable. I have now set myself a massive challenge: to explain how genes can cause the development of behaviour. In this book so far I have had a stab at showing how a pair-bonding instinct is manifest in oxytocin receptor genes, or how personality is affected by BDNF genes. These are useful systems to analyse. But they raise an enormous question: how did the brain get to be built that way in the first place? It is all very well saying that oxytocin receptors expressed in the medial amygdala fire up the dopamine system with sensations of personal addiction towards the loved one. But who built the darned machine this way and how?

Think of the Genome Organising Device as a skilful chef, whose job is to bake a soufflé called the brain. How does it go about this task?

SIGNPOSTS IN THE MIND

Consider, first, the sense of smell. At the perceptual level smell is a genetically determined sense: one gene, one scent. The mouse has 1,036 different olfactory sensors in its nose, each expressing a slightly different olfactory receptor gene. Human beings, in this respect as in many others, are impoverished: they have only 347 intact olfactory receptor genes, plus many rusting hulks of old genes (called pseudogenes).[13] In the mouse, each cell then sends a single nerve fibre (an axon) to a different unit within the olfactory bulb of the brain. Remarkably, the cells expressing one kind of receptor gene all send their axons to just one or two units.

So, for instance, the P2 neurons in the mouse nose – several hundred of them – all express the same receptor gene and supply all of their electrical output to stimulating just two foci in the brain. There is a steady turnover in the neurons, which live for only 90 days. Their replacements grow into the brain and reach exactly the same spot as their predecessors. A team in Richard Axel's laboratory at Columbia University hit upon the devastating idea of killing all the P2 cells (by making them, and only them, express diphtheria toxin) and then seeing if their replacements could still find their way with no 'colleagues' to hold their hands along the way. They could.[14]

This might explain why smells are so evocative. The olfactory neurons are so faithful to the same brain foci that even though the neurons of childhood are long gone, their adult replacements follow exactly the same course into the brain. When Axel and his colleagues removed the odorant receptor gene from P2 cells, they no longer grew to their target, but wandered aimless in the brain. When he replaced the P2 odorant receptor gene with one from P3, the axon now found its way directly to the P3 target.[15] This proves that the development of a specific sense of smell requires a gene expressed in the nose, and a gene expressed in the brain that matches it, the axons growing to make the link.

The first insight to explain how this comes about was the work of a rather romantic contemporary of my 12 hairy men. Santiago Ramon y Cajal (1852–1934) was everything that a Spanish hero should be: artistic, flamboyant, restless and athletic. It was Cajal who convinced the world that the brain is made not of a continuous network of interconnected nerve fibres, but of many separated cells, each touching, but not merging with others. He gets slightly more credit for this discovery than he deserves, since it was an insight shared by at least five other scientists, including the Norwegian explorer and statesman Fridtjof Nansen. But Nansen had quite enough to be famous for, so give Cajal his due. However, it was Cajal's other intuition that interests me here. Cajal suggested that the nervous system is built by nerves growing towards chemicals that attract them. He suspected that nerves are lured to their destinations by gradients of some special substance. In this he was absolutely right.

Like one of Macbeth's witches, I must now add to my recipe the eye of a frog. Frogs have binocular vision: they can look forward with two eyes, all the better to do range-finding on passing flies. Tadpoles, however, have eyes on the sides of their head. Since the tadpole grows into a frog, the eyes have to move into their new positions halfway through life. Problem: now the two eyes' fields overlap so they see the same scene. The frog's brain must take the inputs from the left half of each eye and send them to the same part of the brain for processing together. Meanwhile, the right half of the visual field of each eye must be analysed in a different place. To do this, the GOD must change the wiring from the eye to the brain. The nerve cells from one half of each eye must cross over to the contralateral side of the brain, and those from the other half must stay on the same side. Incredibly, thanks to the work of Christine Holt and Shin-ichi Nakagawa, it is possible to describe exactly how this is done.[16]

Each cell in the retina of the eye grows an axon towards the 'optic tectum' of the brain. At the tip of the axon is an object called a 'growth cone', which seems to be a sort of locomotive for the axon, capable of pulling the tip of the axon in a straight line, or turning or stopping. It does each of these manoeuvres in response to chemicals that attract and repel it. When the growth cones from the tadpole's eye reach the optic chiasm, a sort of crossroads or points junction, they cross over each other so that the right half of the tadpole's brain responds to the left eye and vice versa. But once the tadpole starts to become a frog, something changes at the chiasm. Now the nerves from the left half of the right eye and the left half of the left eye must end up in the same place, and the right halves in another place, so that the frog can see in stereo, the better to judge the distance of passing flies. New neurons grow from each retina to the brain, but this time, half of them cross over the chiasm, while the other half continue into the same side of the brain. Holt and Nakagawa discovered how this change is effected. A gene is switched on within the chiasm: the gene for a protein called ephrin B, which repels the growth cones. It only repels the growth cones coming from one half of each eye, because only half the cells are expressing the gene for the ephrin B receptor. The repelled cones continue into the same side of the brain as the eye they came from. The cells from the other half of the eye, not expressing the receptor, ignore the signal from ephrin B and cross to the contralateral side of the brain. The effect is to give the frog binocular vision so it can range-find flies.

Using just two genes – ephrin B and the ephrin B receptor – expressed in the right pattern in the right places at the right times, the frog has acquired the wiring that enables it to see in a binocular fashion. Exactly the same genes are expressed in exactly the equivalent places in a foetal mouse, whereas in a fish or a chick the genes remain silent and no binocular vision is achieved – which is just as well, since fish and chickens have eyes on the sides of their heads, not the fronts of their heads.

Ephrin B is an 'axon guide', one of a surprisingly small number of such proteins. There are four common families of axon guidance proteins: netrins, ephrins, semaphorins and slits. Netrins generally attract axons, while the others generally repel them. Some other molecules also act as axon guides, but the number is not large. Yet it is beginning to look as if these happy few are almost all that are needed in brain building, because the same four kinds of axon guides are cropping up wherever scientists look, repelling or attracting growth cones

– and in almost all animals, including the lowliest worms. It is a system of mind-boggling simplicity, yet it seems to be capable of producing a human brain with a trillion neurons, each making a thousand connections.[17]

Indulge me in one more case history from the molecular biology of axon guidance before I let you climb back up into psychology for air. In fruit flies, as in frogs, some axons are required to cross the mid-line of the animal to the other side of the brain. To do so, they need to suppress their sensitivity to 'Slit', a repulsive axon guide stationed at the mid-line. An axon that wishes to cross the mid-line must suppress its expression of a gene called Robo, which encodes the receptor for Slit. This suppression makes the axon insensitive to Slit, allowing it free passage through the mid-line checkpoint. Once the axon has crossed, Robo switches back on, which prevents recrossing. The axon may then switch off extra Robo genes (called Robo2 and Robo3), which determine how far from the mid-line it will go. The more Robos it switches off, the farther from the mid-line it will travel.

Although these genes were found in flies, it was no surprise when a mutant zebra fish soon turned up with the exact equivalent of the Robo3 gene not working and with problems at the mid-line nerve crossings. Then came three Slits and two Robos in mice, again doing exactly the same job, directing traffic at the mid-line during the formation of the forebrain. In mice, however, the Slits may do more: they may actually channel axons towards particular regions of the brain.[18] It appears that Slit and Robo genes keep switching on and off in different parts of the rodent brain long after birth, guiding axons to their destinations.[19] Since, with respect to such genes, people are just big mice, this looks like a real breakthrough in understanding how human mental networks get built.

You may think this is a long way from behaviour, and it surely is. My purpose so far is merely to show in outline how genes might set about building a brain according to a very complicated recipe but one that employs a few simple rules – and to show the fourth dimension of genetics, the dimension of time. I do not mean to imply that brain development is now fully understood and scientists are just filling in the details. Far from it. As always in science, the more that a scientist knows, the more that he realises he does not know. Until now fog hid the view before us. All that has happened is that it has partly dissolved to reveal glimpses of a giddy abyss of ignorance. I cannot begin to tell you how netrin and ephrin are affected by experience, for example, nor how a cuckoo's brain is equipped by these axon guides with the instinct to sing 'cuckoo'. But a start has been made. And I cannot resist pointing out that this beginning has come about through genetic reductionism. To try to understand the construction of the mind without considering the individual genes involved in axon guidance would be like trying to create a forest without planting any trees.

EX UNUM PLURIBUS

The axon guides, standing at their guideposts directing the passing growth cones according to their receptors, are only part of the story. They explain how nerves get where they want to go, but they cannot explain how nerves make the right connections when they get there. Parable time again. Suppose a woman from London is offered a job trading bonds in New York. She migrates to New York by responding to certain signals at guideposts along the way (the railway station, the terminal, the check-in desk, the gate, the arrivals hall, the taxi rank, the hotel, the subway, and so on) until she reaches the offices of her new employer. Here, suddenly, she switches to a different kind of navigating: she connects with her new boss and her future colleagues, some of whom have also travelled from afar to that office. She finds them not by directional cues, but by personal ones – name and job. In much the same way, the GOD, having guided an axon to its destination, must connect it with appropriate other neurons on arrival. The cues are no longer directional signs, but badges of identity.

In the late 1980s scientists chanced upon the first example of a gene that tells a migrating axon when it has reached its destination.

The story begins in 1856, when a Spanish doctor named Aureliano Maestre de San Juan carried out a post-mortem on a 40-year-old man who had no sense of smell, a small penis and very small testes. In the man's brain San Juan could find no olfactory bulbs. A few years later another case turned up in Austria, and doctors began to ask men with minute penises if they had a sense of smell. Excitable sexologists took these cases as evidence that noses and penises had as much in common as met the eye. In 1944, Franz Kallmann, a psychologist I mentioned in chapter 4, described the syndrome of small gonads and no sense of smell as a rare genetic disorder, running in families but affecting mainly men. Somewhat unfairly, the syndrome is now named after Kallmann and not the polynomial Spaniard: that's what you get for having so many names.

The search for the genes involved in Kallmann syndrome zeroed in on the X chromosome (of which men have no spare copy because they inherit it from the mother only) and soon pinned it down to a gene called KAL-1. There are almost certainly two other genes on other chromosomes that can also cause Kallmann syndrome, but they remain unidentified. In recent years, it has become clear how KAL-1 works and what happens when it is broken. The gene is switched on about five weeks after conception in neither the nose, nor the gonads, but in the part of the embryonic brain that will become the olfactory bulb. It produces a protein called anosmin, which acts as a cell-adhesion molecule – that is, it causes cells to stick to each other. Anosmin somehow has a dramatic effect on the growth cones of migrating olfactory axons heading for the olfactory bulb. As these growth cones arrive at the brain in the sixth week of life, the presence of anosmin causes them to expand and to 'defasciculate', a posh word for derail. Each axon leaves its tracks and stops, connecting with the cells nearby. In people who have no working copy of KAL-1, and no anosmin, the axons never make a connection with the olfactory bulb. Feeling unwanted, they wither away.[20]

Hence the lack of a sense of smell in people who have Kallmann syndrome. But why the small penis? Astonishingly, it appears that the cells necessary for triggering sexual development also begin life in the nose, in an ancient pheromone receptor called the vomeronasal organ. Unlike the olfactory neurons, which merely send axons to the brain, these neurons themselves migrate to the brain. They do so along the fascicules – the rails – already laid down by the olfactory axons. In the absence of anosmin, they never reach their target, and never begin their main task: the secretion of a hormone called gonadotropin-releasing hormone. Without this hormone, the pituitary gland never gets its instruction to start releasing luteinising hormone into the blood, and without luteinising hormone the gonads never mature, the man has low testosterone levels and therefore low libido, and he remains sexually indifferent to women even after puberty.[21]

Hooray! At last I have found a way to trace the pathway from a gene to a behaviour via the building of a part of the brain. Pat Bateson cites Kallmann syndrome to stress that though genes can indeed influence behaviour, the connections are tortuous and indirect. To call KAL-1 'the gene for' sexual dysfunction would be misleading, not least because it only creates the dysfunction when not working. Besides, anosmin probably has several other functions in the body. Its effect on sexual development is indirect. And there are several other genes that can go wrong that cause some or all of the same symptoms and that are probably working at other points along the extended sequence of causes and effects. Indeed, the majority of inherited Kallmann cases are caused by mutations in other genes than KAL-1.[22]

Although there is no one-to-one correspondence between genes and behaviour (but many-to-many), nevertheless KAL-1 is still, in a cautious and accidental sense, 'one of the genes for' part of sexual behaviour. Just as Lehrman and Piaget might have argued, it manifests its behavioural effect via the physical development of the nervous system. The gene specifies how development occurs and that in turn specifies how behaviour occurs. The spooky truth is dawning on scientists that they can regard behaviour as just an extreme form of development. The nest of a bird is just as much a product of its genes as its wings are. In my garden and all over Britain song thrushes line their nests with mud, blackbirds with grass, robins with hair and chaffinches with feathers, generation after generation, because nest building is an expression of genes. Richard Dawkins coined the phrase 'the extended phenotype' for this idea.[23]

I mentioned that anosmin is a cell-adhesion molecule, and this makes it one of the most intriguing items in the GOD's portfolio of gene products. It is early days yet in understanding the role cell-adhesion molecules play, but it is looking increasingly plausible that they are the badges by which neurons identify their team colleagues when the brain is being wired. They are the key to cells finding each other in the crowd. I justify this highly speculative assertion on the basis of the following experiment, probably the most ingenious I have yet encountered in the study of genes and brains.

The impresario of the experiment is Larry Zipursky; the subject is a simple fruit fly. Flies have compound eyes – that is, their eyes are divided into 6,400 little hexagonal tubes, each focused on one small part of the scene. Each of these tubes sends precisely eight axons to the brain to report on what it sees – mainly movement. Six of these axons respond best to green light; the seventh reports ultraviolet light; the eighth responds to blue light. The first six stop at an early layer of the brain; the seventh and eighth penetrate deeper, the seventh going deepest into the brain.[24] Zipursky first showed that almost certainly for all eight of these cells to reach their targets the gene for N-cadherin (a cell-adhesion protein) must be switched on in the eight cells and also in their targets. What his team then did, almost incredibly, was to genetically engineer a fly so that a few of the seventh cells only express a mutant version of the N-cadherin gene, and they, and only they, turn fluorescent green, allowing the experimenter to distinguish between the development of a mutant and normal cell in the same animal. The details of how this is achieved are mind-boggling: they show that science is still a place for ingenuity and virtuosity. Without N-cadherin, the seventh axon develops normally, reaches its target, but then fails to make a connection, retracts and seems to become disorientated. Zipursky repeated the experiment with the first six neurons and they too could not find their destination without the N-cadherin gene working. He concludes that N-cadherin (and, after a similar experiment, another gene called LAR, also a cell-adhesion gene) is necessary for an axon to recognise its target in the brain.[25]

Cadherins and their ilk are currently among the most glamorous molecules in biology. They owe this reputation to the role they are thought to play in enabling neurons to find each other during the wiring of the brain. They stick out of the surface of neurons like fronds of kelp from the seabed. In the presence of calcium, they stiffen into rods and grab hold of similar cadherins from neighbouring cells. Their job seems to be to bind together two neurons. But they will only bind to each other if their tips are compatible, and the Genome Organising Device seems to go to great lengths to vary the tip of the frond between different cells. This is partly because there are many different cadherin genes, but it is partly due to an entirely different phenomenon named alternative splicing. Bear with me while I take you on a tour of the workings of genes. A gene is a stretch of DNA letters encoding the recipe for a protein. In most cases, however, the gene is broken up into several short stretches of 'sense' interrupted by long stretches of nonsense. The sense bits are called exons and the nonsense bits introns. After the gene has been transcribed into a working copy made of RNA and before it has been translated into protein, the introns are removed in a process called splicing.

This was discovered in 1977 by Richard Roberts and Philip Sharp and gained them a Nobel Prize. Walter Gilbert then realised that there was more to splicing than merely cutting out the nonsense. In some genes, there are several alternative versions of each exon, lying nose to tail, and only one is chosen; the others are left out. Depending on which one is chosen, slightly different proteins can be produced from the same gene. Only in recent years, however, has the full significance of this discovery dawned. Alternative splicing is not a rare or occasional event. It seems to occur in approximately half of all human genes;[26] it can even involve the splicing in of exons from other genes; and in some cases it produces not just one or two variants from the same gene, but hundreds or even thousands.

In February 2000, Larry Zipursky had asked one of his graduate students, Huidy Shu, to look at a molecule called Dscam, a gene product recently purified in the fly by Jim Clemens and shown by Dietmar Schmucker to be required for guiding fruit fly neurons to their targets in the brain. One part of the fly gene looked disappointingly different in one small region from its human equivalent, a gene that probably causes some of the symptoms of Down syndrome by an unknown mechanism (Dscam stands for Down syndrome cell-adhesion molecule). Shu began looking for alternative forms of Dscam that might contain regions of sequence similar to the human gene; and while no such sequence was identified, surprisingly every one of the 30 or so forms of Dscam that Shu sequenced was different. Then suddenly, for the first time, the entire fruit fly genome became available over the internet from the Celera corporation. That weekend Shu and Clemens used the database to read the Dscam gene. They could not believe their eyes when the result of the search came through. There were not a few alternative exons; there were 95. Of the 24 exons in the gene, four existed in alternative versions: exon 4 came in 12 different versions, exon 6 in 48, exon 9 in 33 and exon 17 in 2. That meant that, if the gene were to be spliced into every possible combination of exons, it could produce 38,016 different kinds of protein – from one gene![27]

News of the Dscam discovery spread like wildfire through the community of geneticists. Many genome experts found it rather depressing, for it suddenly threw an immense spanner of complexity into the works. If a single gene could make thousands of proteins, then listing human genes would only be the very beginning of the task of listing the number of proteins they could produce. On the other hand, such complexity made a mockery of the argument that the comparatively few genes in the human genome meant the genome was too simple to explain human nature, so people must be the product of experience instead. Those who argued this were suddenly hoist by their own petard. Having argued that a genome of 30,000 genes was too small to determine the details of human nature, they would have to admit that a genome that could produce hundreds of thousands, perhaps even millions, of different proteins had easily enough combinatorial capacity to specify human nature in excruciating detail, without even bothering to use nurture.

It is important not to get carried away. Few other alternatively spliced genes show such potential diversity. At the time of writing no human version of Dscam (and there are several) has yet proved to be alternatively spliced at all, let alone to such a degree. Nor is it yet known that fruit flies make all 38,016 proteins that they could from Dscam. It remains possible that all 48 versions of exon 6 are functionally interchangeable. But Zipursky already knows that different alternatives of exon 9 are found preferentially in different tissues, and he suspects the same may be true of the other exons. There is a pervasive feeling among the scientists working on this topic that they are scratching at the door of a chamber of secrets. How genes splice themselves and how RNA behaves in the cell may hold the key to some truly fundamental new biological principles.

In any case, Zipursky hopes he may have hit upon a molecular basis for cell recognition: for how neurons find each other in the crowded brain. Dscam is similar in structure to an immunoglobulin, a highly variable protein used in the immune system to identify many different pathogens. Recognising pathogens might be a rather similar business to recognising neurons in the brain.[28] Cadherins and another kind of cell-adhesion molecule used in the brain – protocadherins – also exhibit immunoglobulin-like features. They use alternative splicing that would enable them to have highly specified identity badges. Moreover, the proteins they produce all stick out of cells, waving their variable tails, and stick to each other by matching those tails. Once stuck together with a similar protein from another cell, the tails form a rigid bridge. This looks increasingly like a system whereby like finds like: cells that express the same exons can bind together and form synaptic connections.

In particular, the protocadherins look highly intriguing. Their genes are arranged, head to tail, in three clusters on human chromosome 5, nearly 60 genes in all. Each gene contains a string of variable exons from which to choose, and each exon is controlled by a separate promoter.[29] They may even rearrange their genetic message by alternative splicing not within one gene transcript but between different gene transcripts. This gives the brain potentially not thousands of different protocadherins, but billions. Neighbouring cells in the brain of very similar types end up expressing slightly different protocadherins. 'Protocadherins may therefore provide the adhesive diversity and molecular code for specifying neuronal connections in the brain,' write two of their champions at Harvard.[30]

More than 40 years ago a neuroscientist named Roger Sperry set out to topple the prevailing consensus, championed by his own supervisor, that the brain was created by learning and experience from an undifferentiated, almost random network of neurons. On the contrary, he found that a nerve gets its identity early in development and cannot easily be reprogrammed. By severing and regenerating nerves in salamanders, he proved that each neuron finds its way to the same place as its predecessor. By rewiring the brains of rats and frogs he proved that there was a limit to the plasticity of the animal mind: a rat rewired so that its right foot was now connected with the nerves from its left would continue to move its left foot if the right foot was stimulated. By stressing the determinism in the nervous system, Sperry brought about a nativist revolution in neuroscience that paralleled Chomsky's in psychology. He even postulated that each neuron would have a chemical affinity for its target and the brain would prove to be built by a large number of variable recognition molecules. In this he was far ahead of his time (his Nobel Prize was for other, lesser work).

NEW NEURONS

The story of development, then, seems at first to lead to a rather different conclusion from that which Piaget and Lehrman expected. Just as the study of twins was expected to reveal a large role for the environment and a small role for genes, but found the opposite, so development seems to be a rather well-determined process planned and plotted by genes. Am I to conclude that nature wins this particular argument and that the developmentalist challenge therefore fails?

No. For one thing, a deterministically constructed machine can still be modified. My computer has exquisitely specified circuitry, but that does not stop it modifying the activity of its connections in response to a new program. Besides, neural plasticity is back in fashion since Sperry's day. This is partly because of the usual nature–nurture rebound: today's scientists are reacting to what they see as excessive nativism just as Sperry was reacting to what he saw as excessive empiricism. But there is more to it than that. For many years it was orthodoxy, apparently proved by the neuroscientist Pasco Rakic, that animals grew no new neurons in the cortex of the brain after reaching adulthood. Then Fernando Nottebohm found that canaries make new neurons, when they learn new songs. So Rakic said mammals grow no new neurons, whatever birds do. Then Elizabeth Gould found that rats do. So Rakic retreated to primates. Gould found new neurons in tree shrews. So Rakic said it was higher primates. Gould found them in marmosets. So it was Old World primates. Gould found them in macaques. Now it is certain that all primates, including human beings, can grow new cortical neurons in response to rich experiences, and lose neurons in response to neglect.[31] There is ample and growing evidence that, for all the determinism in the initial wiring of the brain, experience is essential for refining that wiring. In Kallmann syndrome, the olfactory bulbs wither away for lack of use. The old public accounting principle for how to handle a government grant – 'use it or lose it' – seems to hold sway in the mind as well.

Notice a tendency to accentuate the negative. The best way to prove the importance of experience is to deprive an animal of it. In the visual cortex, an eye blindfolded at birth soon loses its receptive field in the brain to the other eye (more on this in the next chapter). However, as I write, Hollis Cline has just produced the first experimental evidence of how experience positively affects the development of the brain. She studies the way a neuron from the eye behaves when it nears its target in the brain. Far from homing in on its goal in a predetermined way, it throws out a whole 'arbor' of feelers, many of which are soon retracted. It seems to be looking for connections that 'work' – connections between like-minded neurons that fire together. Cline compared neurons in the visual system of a developing tadpole after four hours of light stimulation or four hours of dark and showed that the cell had thrown out far more feelers looking for contacts in the light. 'I've got a stimulus,' the neuron cries, 'I want to share the news.' This may be how experience actually affects the development of the brain, just as Piaget argued. Cline's colleague, Karel Svoboda, has actually watched through a window in the skull as synapses between the brain cells of a mouse form and dissolve in response to experience.[32]

The whole point of education is surely to exercise those brain circuits that might be needed in life – rather than to stuff the mind full of facts. Thus exercised, they flourish. Astonishingly, this is something human beings share with microscopic worms. The nematode worm *Caenorhabditis elegans* is the reductionist's delight. It has no brain and exactly 302 neurons – wired up according to a rigid programme. It seems like one of the least likely candidates for even the simplest form of learning, let alone developmental plasticity and social behaviour. Its behaviour consists of not much more than wriggling forwards and wriggling backwards. Yet if such a worm repeatedly finds food at a certain temperature, it registers this fact and thenceforth shows a preference for this temperature; if unrewarded at this temperature, it gradually loses its temperature preference. Such flexible learning is under the influence of a gene called NCS-1.[33]

Not only can nematode worms learn, they can also develop different adult 'personalities' according to their social experience during infancy. Cathy Rankin sent some worms to school (i.e., reared them together in a single Petri dish) and kept others at home (i.e., alone in a dish). She then tapped the side of each dish, which caused the worms to reverse the direction of their movement. The social worms, which were used to running into each other, were much more sensitive to the tapping than the solitary worms.

Rankin had engineered certain genes inside the worm so that she could study exactly which synapses between which neurons were responsible for the difference between the social and the solitary worms. The differences showed up as weaker glutamate synapses between certain sensory neurons and 'interneurons'. Intriguingly, she found that the very same synapses could be altered during learning. After 80 taps, worms of both kinds became habituated to the fact that they lived in a vibrating world and gradually lost their tendency to reverse direction: they had learned. Both learning and schooling exerted their effects at the same synapses and they did so by altering the expression of the same genes.[34]

To prove that the development of behaviour in a humble worm is environmentally plastic in this way rather underlines the developmentalist challenge. If an organism with no brain and just 302 neurons can benefit from going to school, then how much greater will be the effect of such contingencies in human upbringing. It is abundantly clear that early social enrichment has long-lasting and irreversible effects on the behaviour of mammals. In the 1950s Harry Harlow (of whom, more in chapter 7) discovered accidentally that a female monkey reared in an empty cage with just a wire model of a mother for company and no peers to play with will grow up to be a neglectful mother herself. She treats her babies as if they were large fleas. She has been somehow imprinted with the impoverished experience of her childhood and passes it on.[35]

Likewise, baby mice separated from their mothers, or handled by human beings, are permanently affected by the experience. Isolated offspring grow up to be anxious, aggressive and slightly more vulnerable to drug addiction. A mouse that was licked a lot by its mother as a baby tends to lick its own pups a lot, and cross-fostering reveals that this is inherited non-genetically – an adopted mouse will behave more like its nursing than its biological mother. There is little doubt that these effects are mediated through genes in the baby mouse.[36]

A female mouse presented with pups will ignore them at first but will gradually become maternal towards them. The speed with which this response occurs varies greatly between mice, and again a mouse that was licked a lot as a baby will respond more quickly. The work of Michael Meaney suggests that the genes involved are those for oxytocin receptors, which are switched on more easily in the mice that were well licked as babies. Somehow, the early licking alters the sensitivity of these genes to oestrogens. Quite how this works is not known, but it may involve the dopamine system of the brain, dopamine being a mimic of oestrogen. The plot thickens, because early maternal neglect definitely changes the expression of genes involved in the development of the dopamine system, which apparently accounts for the fact that animals from deprived backgrounds are more easily addicted to certain drugs – drugs reward the mind through the dopamine system.[37]

Darlene Francis in Tom Insel's lab took two strains of mice and swapped them before and after birth. Mice of the C57 strain, transplanted just after fertilisation, were nurtured in the wombs of mice of either their own strain or the BALB strain and then reared either by BALB or C57 mothers. After all this cross-fostering, the mice were tested for their skills at various standard tests which all mice living in laboratories are habitually required to take. One test involves finding a hidden platform on which to stand in a milky swimming pool and then remembering where it is. Another test involves plucking up the courage to explore when dropped in the middle of an open space. A third test involves exploring a cross-shaped maze in which two of the arms are closed and two open. The inbred strains of mice consistently differ in their performance on these tests, implying that genes prescribe their behaviour. BALB mice spend less time in the middle of the open field, spend more time in the closed arms of the cross and recall faster where to find the hidden platform than C57 mice. In the cross-fostering experiment, the C57 mice cross-fostered to C57 mothers either before or after birth behaved just like normal C57 mice. But C57 mice cross-fostered to BALB mothers just after fertilisation and then reared by BALB mothers behaved just like BALB mice. Like Meaney's rats, the BALB mothers lick their pups less than the C57 mothers, and seem thereby to change the pups' natures. But this effect of maternal behaviour depends on growing up in a BALB womb. C57 pups from a C57 womb that are cross-fostered to a BALB mother after birth look just like other C57 mice and not at all like BALB mice. As Insel puts it, Mother Nature meets Mother Nurture.[38]

These are stunning discoveries. They hint at enormous sensitivity in the development of the mammal brain to how its owner is treated in the womb and soon after birth, but they also suggest that these effects are mediated through the animal's genes. It is a striking example of Lehrman's point that development matters to the adult outcome. Indeed, it goes further than Lehrman did in revealing how genes are at the mercy of the behaviour of other animals in the environment, especially parents. As usual, it supports neither an extreme nurture argument (because it is a phenomenon made possible by the actions of genes), nor an extreme nature argument (because it shows how plastic the expression of genes can be). It reinforces my message that genes are servants of nurture as much as they are servants of nature. It is a beautiful example of how the GOD includes in the job description of some genes the following admonition: during development you should at all times be ready to absorb information from the environment outside your parent organism and adjust your activity accordingly.

INCUBATING UTOPIA

'Hasn't it ever occurred to you that an Epsilon embryo must have an Epsilon environment as well as Epsilon heredity?' So speaks the Director of Hatcheries and Conditioning in Aldous Huxley's novel of 1932, *Brave New World*. He is showing students the Predestination and Decanting Rooms in the hatchery, where artificially inseminated human embryos are reared in different conditions to produce different castes of society: from brilliant alphas to factory-fodder epsilons.

Rarely has a book been more misrepresented than *Brave New World*. It is today almost automatically assumed to be a satire on extreme hereditarian science: an attack on nature. In fact it is all about nurture. In Huxley's imagined future, human embryos, having been artificially inseminated and in some cases cloned ('Bokanovskified'), are then developed into members of the various castes by a careful regime of nutrients, drugs and rationed oxygen. This is followed, during childhood, by incessant hypnopaedia (brainwashing during sleep) and neo-Pavlovian conditioning until each person emerges certain to enjoy the life to which he or she has been assigned. Those who work in the tropics are conditioned to heat; those who fly rocket planes are conditioned to motion.

The highly 'pneumatic' heroine Lenina is predestined – by what was done to her in the hatchery and in school, not by her genes – to enjoy flying, dates with the assistant predestinator, casual sex, rounds of obstacle golf and doses of the happiness drug, Soma. Her admirer, Marx, rebels against such conformity only because alcohol was mistakenly added to his blood-surrogate before birth. He takes Lenina to a Savage Reservation in New Mexico for a holiday, where they meet Linda, a white 'Savage', and her son, John, whom they bring back to London to confront John's father, who turns out to be the director of hatcheries and conditioning himself. John, autodidactically educated by a volume of Shakespeare, longs to see the civilised world, but becomes rapidly disillusioned with it and retires to a lighthouse in Surrey, where he is tracked down by a film-maker. Goaded by intrusive spectators, he hangs himself.[39]

Although there are drugs to keep people happy, and hints of heredity, the details of *Brave New World*, and the features that make it such a horrific place to live, are the environmental influences exercised upon the development of the bodies and brains of the inhabitants. It is a nurture hell, not a nature hell.

C H A P T E R S I X

Formative years

The childhood shows the man, as morning shows the day.
John Milton, *Paradise Regained*[1]

Nurture is reversible; nature is not. That is the reason responsible intellectuals have spent a century preferring the cheerful meliorism of environment to the bleak Calvinism of genes. But what if there were a planet where it was the other way round? Suppose some scientist discovered a world in which lived intelligent creatures whose nurture was something they could do nothing about, whereas their genes were exquisitely sensitive to the world in which they lived.

Suppose no more. In this chapter I aim to start convincing you that you live on precisely such an inverted planet. To the extent that people are products of nurture, in the narrowly parental sense of the word, they are largely the products of early and irreversible events. To the extent that they are the product of genes, they are expressing new effects right into adulthood, and often those effects are at the mercy of the way they live their lives. This is one of those contrarian surprises that science delights in delivering, and it is one of the least recognised and most significant discoveries of recent years. Even its discoverers, steeped as they are in the litany of nature versus nurture, are only dimly aware of how revolutionary their discoveries are.

In 1909, in the Danube marshes near Altenberg in eastern Austria, a six-year-old boy named Konrad and his friend Gretl were given two new-hatched ducklings by a neighbour. The ducklings became imprinted on the children and followed them everywhere, mistaking them for parents. 'What we didn't notice,' said Konrad 64 years later, 'is that I got imprinted on the ducks in the process . . . A lifelong endeavour is fixed by one decisive experience in early youth.'[2] In 1935 Konrad Lorenz, by then married to Gretl, described rather more scientifically how a gosling, soon after hatching, will fixate on the first moving thing it encounters and follow it. That moving thing is usually its mother, but occasionally it turns out to be a goatee-bearded professor. Lorenz realised that there was a narrow window of time during which this imprinting could occur. If the gosling was less than 15 hours or more than three days old, it would not imprint. Once imprinted, it was stuck and could not learn to follow a different foster-parent.[3]

Lorenz was not actually the first to describe imprinting. More than 60 years before, the English naturalist Douglas Alexander Spalding spoke of early experience being 'stamped in' to a young animal's mind – virtually the same metaphor. Little is known about Spalding, but that little is refreshingly exotic. John Stuart Mill, having met Spalding in Avignon, got him the job of tutor to the elder brother of Bertrand Russell. Russell's parents, Viscount and Viscountess Amberley, thought it would be wrong for Spalding, a consumptive, to reproduce. But they thought it equally wrong that a man's natural sexual urges should be denied, so they decided the dilemma should be solved in the obvious way: by Lady Amberley personally. Dutifully she did so, but in 1874, she died, followed in 1876 by her husband, who had named Spalding as one of Bertrand Russell's guardians. The revelation of the affair appalled the aged grandfather, Earl Russell, who promptly took over the guardianship of young Bertrand before himself dying in 1878. Spalding, meanwhile, had died in 1877 of his tuberculosis.

The obscure hero of this Greek tragedy seems in his few writings to have anticipated many of the great themes of twentieth-century psychology, including behaviourism. He also described how a newborn chick 'will follow any moving object. And, when guided by sight alone, they seem to have no more disposition to follow a hen than to follow a duck or a human being . . . there is the instinct to follow; and the ear, prior to experience, attaches them to the right object.' Spalding even remarked on how a chick kept hooded for the first four days of life immediately fled from him when unhooded, whereas if it had been unhooded the day before, it would have run to him.[4]

But Spalding went unnoticed and it was Lorenz who put imprinting (in German, 'Pragung') on the scientific map. It was Lorenz who coined the concept of the critical period – the window during which environment acts irreversibly upon the development of behaviour. For Lorenz the importance of imprinting was that it was itself an instinct. The tendency to imprint on a parent is innate in the new-hatched gosling. It cannot possibly be learned, for it is the bird's first experience. At a time when the study of behaviour was dominated by conditioned reflexes and associations, Lorenz saw his role as rehabilitating innateness. In 1937 Niko Tinbergen spent the spring with Lorenz at Altenberg and between them they invented the science of ethology – the study of animal instincts. Concepts like displacement (doing something else when prevented from doing what is desired), releasers (the environmental triggers of instinct) and fixed action patterns (subprograms of an instinct) were born. Tinbergen and Lorenz were awarded the Nobel Prize in 1973 for the work that began that spring.

But there is another way to view imprinting, as the product of the environment. After all, the gosling will not follow unless there is something to follow. Once it has followed one kind of 'mother' it will prefer to follow one that looks like that. But before then, it is open-minded about what 'mother' looks like. From a different perspective, Lorenz had discovered how the external environment shapes behaviour just as much as the internal drive does. Imprinting could be recruited to the nurture camp as surely as it was recruited to the nature camp: a gosling can be taught to follow anything that moves.[5]

A duckling, however, is different. Despite his boyhood success with ducklings, the adult Lorenz could not easily get mallard ducklings to imprint upon him until he tried making mallard-like noises. Then they followed him with enthusiasm. The ducklings need both to see and to hear their mother. In the early 1960s, Gilbert Gottlieb did a series of experiments to explore how this works. He found that naïve newborn ducklings of either mallard or wood ducks had a preference for the calls of their own species. That is, despite never having heard their own species call, they knew the right sound when they heard it. But Gottlieb then tried to complicate things and got a surprising result. He muted the ducklings themselves by operating on their vocal cords while they were still in the egg. Now the ducklings, on hatching, had no preference for their own species of mother. Gottlieb concluded that the ducklings only knew the right call because they had heard their own voices before hatching. This he felt undermined the whole notion of instinct, by bringing an environmental trigger in before birth.[6]

THE SCARS OF GESTATION

If the influence of the environment is partly prenatal, then the environment begins to sound a lot less like a malleable force and more like fate. Is this a curiosity of ducks and geese, or are people also imprinted by the early environment with certain unvarying characteristics? Start with the medical clues. In 1989, a medical scientist named David Barker analysed the fate of more than 5,600 men born between 1911 and 1930 in six districts of Hertfordshire in southern England. Those who had weighed the least at birth and at one year old went on to have the highest death rates from ischaemic heart disease. The risk of death was nearly three times as great in the light babies as in the heavy babies.[7]

Barker's result gathered much attention. It was no surprise that heavier babies should be more healthy, but it was a great surprise that they should be less vulnerable to a disease of old age, and one, moreover, for which the causes were supposedly well known. Here was evidence that heart disease is not so much influenced by how much cream you eat as an adult, but how thin you were at one year old. Barker has gone on to confirm the same result in data from other parts of the world for heart disease, stroke and diabetes. For instance, in 4,600 men born in Helsinki University Hospital between 1934 and 1944, those who were thin or light at birth and one year old were far more likely to die of coronary heart disease. Barker puts it this way: had none of these people been thin as babies, then there would have been half as much coronary heart disease later – a huge potential gain to public health.

Barker argues that heart disease cannot be understood as an accumulation of environmental effects during life. 'Rather, the consequences of some influences, including a high body mass in childhood, depend on events at early critical stages of development. This embodies the concept of developmental "switches" triggered by the environment.'[8] According to the 'thrifty phenotype' hypothesis, which has grown out of this work, Barker has found an adaptation to famine. A poorly nourished baby's body, imprinted with its prenatal experience, is born 'expecting' to live in a state of food deprivation throughout its life. Its whole metabolism is geared to being small, hoarding calories and avoiding excessive exercise. When, instead, the baby finds itself in a time of plenty, it compensates by growing fast but in such a way as to put a strain on its heart.

The famine hypothesis may have even more bizarre implications, as revealed by an 'accidental experiment' conducted on a vast scale during the Second World War. It began in September 1944, at a time when the former collaborators Konrad Lorenz and Niko Tinbergen were both in captivity. Lorenz was in a Russian prisoner-of-war camp, having just been captured; Tinbergen was about to be released after two years in a German internment camp held hostage under threat of death against the activities of the Dutch resistance. On 17 September 1944, British paratroopers occupied the Dutch city of Arnhem to capture a strategic bridge over the Rhine. Eight days later, the Germans forced them to surrender, having fought off the ground forces sent to their relief. The Allies then abandoned attempts to liberate Holland until after the winter.

The Dutch railway workers had called a strike to try to prevent German reinforcements reaching Arnhem. In retaliation, Reichskommissar Arthur Seyss-Inquart ordered an embargo on all civilian transport in the country. The result was a devastating famine, which lasted for seven months: the hunger winter, they called it. More than 10,000 people starved to death. But what later caught the attention of medical researchers was the effect that this abrupt famine had on unborn babies. Some 40,000 people were foetuses during the famine, and their birthweight and later health are all on record. In the 1960s a team from Columbia University studied the data. They found all the expected effects of malnourished mothers: malformed babies, high infant death rates and high rates of stillbirth. But they also found that those babies who were in their last trimester of gestation (only) suffered from low birth weight. These babies grew up normal but they later suffered from diabetes, probably brought on by the mismatch between their thrifty phenotype and the abundant rich food of the post-war world.

Babies who were in the first six months of gestation during the famine were normal in birth weight, but when they reached adulthood they themselves gave birth to unusually small babies. This bizarre second-generation effect is hard to explain with the thrifty phenotype hypothesis, though Pat Bateson notes that locusts take several generations to switch from the shy, solitary form with a specialist diet to the swarming, gregarious form with a generalist diet and back again. If it takes several generations for humans to switch between thrifty and affluent phenotypes, this may explain why Finland has nearly four times the death rate from heart disease as France. The government of France began supplementing the rations of pregnant mothers after the Franco-Prussian war of the 1870s. The people of Finland lived in comparative poverty until 50 years ago. Perhaps it is the first two generations to experience abundance who suffer from heart disease.

Perhaps that is why the United States is now seeing rapidly falling death rates from heart disease, but Britain, well fed for a shorter time, is lagging behind.[9]

THE LONG FINGER OF LIFE

A prenatal event may have far-reaching effects that are all but impossible to counteract in later life. Even subtle differences between healthy individuals can be put down to prenatal imprinting. Finger length is a case in point. In most men the ring finger is longer than the index finger. In women the two fingers are usually the same length. John Manning realised that this was an indication of the level of prenatal testosterone to which people had been exposed while in the womb. The more testosterone in your womb experience, the longer your ring fingers. There is a good biological reason for the link. The hox genes that control the growth of the genitalia also control the growth of digits, and a subtle difference in the timing of events in the womb probably leads to subtly different finger lengths.

Manning's ring finger measurements give a crude measure of testosterone exposure before birth: so what? Well, forget palmistry; this is real prediction. Men with unusually long ring fingers (high testosterone) are at greater risk of autism, dyslexia, stammering, immune dysfunction; they also father relatively more sons.[10] Men with unusually short ring fingers are at higher risk of heart disease and infertility. And because male muscles are also partly testosterone-built, Manning was prepared rather rashly to predict on television that among a group of athletes about to run a race, the one with the longest ring finger would win, which he promptly did.[11]

The length of the ring finger, and indeed the fingerprint on it, are imprinted in the womb. They are the products of nurture – for surely the womb is the very embodiment of the word nurture. But that does not make them malleable. The comforting belief that nurture is more malleable than nature relies partly on the fallacy that nurture is what happens after birth and nature is what happens before birth. Fallacy it is.

Perhaps you can now glimpse an explanation of the paradox of chapter 3: that behaviour genetics reveals a role for genes, and a role for unshared environment, but hardly any role for shared environmental influences. The prenatal environment is not shared with siblings (except twins); the experience of gestation is unique to each baby; the insults suffered therein, such as malnutrition or influenza or testosterone, depend on what is happening to the mother at that time, not on what is happening within the whole family. The more prenatal nurture matters, the less post-natal nurture can matter.

SEX AND THE WOMB

There is something rather Freudian about all this imprinting. Old Sigmund believed that the human mind carries the marks of its early experience, and that many of these marks lie buried in the subconscious, but they are still there. Rediscovering them is one of the delights of the psychoanalyst's couch. Freud went on to suggest that by this process of rediscovery, a person could cure himself of various neuroses. A century later there is an unambiguous verdict on this proposal: good diagnosis, terrible therapy. Psychoanalysis is disastrously bad at changing people. That is what makes it so profitable – 'See you next week.' But it is right in its premise that there are such things as 'formative experiences'; that they come very early; and that they are still powerfully present in the adult subconscious. Yet by this same token, if they are still there, and still influential, then they must be hard to reverse. Formative experiences must be unchangeable, if they persist.

Freud may not have been the first person to consider infantile sexual desires, but he was certainly the most influential. In this he was being contrarian. To the detached observer nothing could be more obvious than that sex starts at adolescence. Until the age of about 12, human beings are indifferent to nudity, bored by romance and mildly incredulous about the facts of life. By 20, they are fascinated by sex to an obsessive degree. Something has surely changed. But Freud was convinced that there was something sexual occurring in the mind of the child, even the baby, long before that.

Back to goslings. Lorenz noticed that imprinted goslings (and other birds) not only treated him as a parent, but later became sexually fixated on him as well. They would ignore members of their own species and court human beings. (My sister and I found the same thing when as children we reared a collared dove from hatchling to adult: it fell fanatically in love with my sisters' fingers and toes, probably because it had been fed with fingers from the moment it opened its eyes. It treated my fingers and toes like sexual rivals.) This was rather intriguing because it implied that, at least in birds, the object of a sexual attraction could be fixed from soon after birth and yet simultaneously could consist of almost any living object. A whole series of experiments both in captivity and in the wild has since shown that in many kinds of bird a male chick reared by a foster-mother of a different species does indeed sexually imprint on that other species, and that there exists a critical period during which it picks up this sexual preference.[12]

Might the same be disturbingly true of people? The reassuring answer that most people gave themselves in the twentieth century was that no, people did not have instincts, so this need not arise. But now see what a fine mess this leads you into! If instinct is something so flexible that a goose can become infatuated with a man, then do human beings have a less flexible instinct? Or do they laboriously have to learn what to love? Either way, the human boast that our lack of instinct is what makes us flexible begins to sound a bit hollow.

In any case, it has long been clear from the experiences of homosexual people that human sexual preferences are not only difficult to change, but also fixed from a very early age. Nobody in science now believes that sexual orientation is caused by events in adolescence. Adolescence merely develops a negative that was exposed much earlier. It is clear that to understand why most men are attracted to women while some men are attracted to men you must go much further back into childhood. Perhaps even into the womb.

The 1990s saw a series of studies that revived the idea of homosexuality as a 'biological' rather than a psychological condition, as a destiny rather than a choice. There were studies showing that future homosexuals had different personalities in childhood; studies showing that homosexual men had differences in brain anatomy from heterosexual men; several twin studies showing that homosexuality was highly heritable in Western society; and anecdotal reports from homosexual men to the effect that they had felt 'different' from an early age.[13] On its own none of these studies was overwhelming. But together, and set against decades of proof that aversion therapy, 'treatment' and prejudice entirely failed to 'cure' people of gay instincts, they were emphatically clear. Homosexuality is an early, probably prenatal and irreversible preference. Adolescence simply throws fuel on the fire.[14]

What exactly is homosexuality? It is plainly a whole range of different behavioural characteristics. In some of them gay men seem to be more like women: they are attracted to men, they pay more attention to clothes, they are often more interested in people than, say, football. In other ways, however, they are more like heterosexual men: they buy pornography and seek casual sex, for example. (*Playgirl*'s nude centrefolds of men turned out to appeal mainly to gays, not the intended women.)[15]

People, like all mammals, are naturally female unless masculinised. Female is the 'default sex' (it is the other way round in birds). A single gene, called SRY, on the Y chromosome starts a cascade of events in the developing foetus that leads to the development of masculine appearance and behaviour. If that gene is absent, a female body results. It is therefore reasonable to hypothesise that homosexuality in men results from the partial failure of this prenatal masculinisation process in the brain, though not in the body (see chapter 9).

By far the most reliable discovery about the causes of homosexuality in recent years is Ray Blanchard's theory of the fraternal birth order. In the mid-1990s Blanchard measured the number of elder brothers and sisters of gay men compared with the population average. He found that gay men are more likely to have elder brothers (but not elder sisters) than either gay women or heterosexual men. He has since confirmed this in 14 different samples from many different places. For each extra older brother, a man's probability of being gay rises by one-third. (This does not mean men with many elder brothers are bound to be gay: an increase from, say, 3 per cent of the population to 4 per cent is an increase of one-third.)[16]

Blanchard calculates that at least one gay man in seven, probably more, can attribute his sexual orientation to this fraternal birth order effect.[17] It is not simply birth order, because having elder sisters has no consequence. Something about elder brothers must actually be causing homosexuality in men. He believes the mechanism is in the womb rather than the family. One clue lies in the birth weight of baby boys who will later become homosexual. Normally, a second baby is heavier than a first baby of the same sex. Boys especially are heavier if they are born after one or more sisters. But boys born after one brother are only slightly heavier than first-born boys, and boys born after two or more brothers are usually smaller than first- and second-born boys at birth. By analysing questionnaires to gay and straight men and their parents, Blanchard was able to show that younger brothers who went on to become homosexuals were 170 grams lighter at birth than younger brothers who went on to become heterosexuals.[18] He confirmed the same result – high birth order, low birth weight compared with controls – in a sample of 250 boys (with an average age of seven) who were showing sufficient 'cross-gender' wishes to have been referred to psychiatrists; cross-gender behaviour in childhood is known to predict later homosexuality.[19]

Like Barker, Blanchard believes that conditions in the womb are marking the baby for life. In this case, he argues, something about occupying a womb that has already held other boys occasionally results in reduced birth weight, a larger placenta (presumably in compensation for the difficulty the baby experiences in growing) and a greater probability of homosexuality. That something, he suspects, is a maternal immune reaction. The immune reaction of the mother, primed by the first male foetuses, grows stronger with each male pregnancy. If it is mild it causes only a slight reduction in birth weight; if strong it causes a marked reduction in birth weight and an increased probability of homosexuality.

What could the mother be reacting to? There are several genes expressed only in males, and some are already known to raise an immune reaction in mothers. Some are expressed prenatally in the brain. One intriguing new possibility is a gene called PCDH22, which is on the Y chromosome, is therefore specific to males, and is probably involved in building the brain.[20] It is the recipe for a protocadherin (yes, them again). Could this be the gene that wires the bit of the brain that is peculiar to males? A maternal immune reaction may be sufficient to prevent the wiring of the part of the brain that would eventually encourage a fascination with female bodies.

Clearly not all homosexuality is caused this way. Some of it may be caused directly by genes in the homosexual person without the mediation of the mother's immune reaction. Blanchard's theory may explain why it has proved so hard to pin down the 'gay gene'. The main method for finding such a gene is to compare markers on the chromosomes of homosexual men with those of their heterosexual brothers. But if many gay men have straight elder brothers, then this method would work poorly. Besides, the key genetic difference might be on the mother's chromosomes, where it causes the immune reaction. This might explain why homosexuality looks as though it is inherited through the female line: genes for a stronger maternal immune reaction could appear to be 'gay genes', even though they may not even be expressed in the gay man himself, but only in the mother.

But notice what this does to the nature-versus-nurture argument. If nurture, in the guise of birth order, causes some homosexuality, it does so by causing an immune reaction, which is a process directly mediated by genes. So is this environmental or genetic? It hardly matters, because the absurd distinction between reversible nurture and inevitable nature has now been well and truly buried. Nurture in this case looks just as irreversible as nature, perhaps more so.

Politically, the confusion is even greater. Most homosexuals welcomed the news in the mid-1990s that their sexual orientation looked like it was 'biological'. They wanted it to be a destiny, not a choice, because that would undermine the argument of homophobes that it was a choice and therefore morally questionable. How could it be wrong if it was innate? Their reaction is understandable, but dangerous. A greater tendency to violence is also innate in the human male. That does not make it right. The naturalistic fallacy, that 'ought' can be derived from 'is', is by definition fallacious. To base any moral position on a natural fact, whether one derived from nature or from nurture, is asking for trouble. In my morality, and I hope in yours, some things are bad but natural, like dishonesty and violence; others are good but less natural, like generosity and fidelity.

THROWING SWITCHES IN THE BRAIN

It is easy to infer the existence of critical periods during which the wet cement of character can be set. It is less easy to conceive how they work. What can possibly occur inside a brain to imprint a gosling on to a professor soon after hatching? Even to ask such a question reveals me to be a reductionist, and reductionists are BAD THINGS. We are supposed to glory in the holistic experience, and not try to take it apart. To which I could reply that there is often more beauty, poetry and mystery in the circuit design of a microchip or the workings of a well-made vacuum cleaner than there is in a room full of conceptual art, but I would not want to be called a philistine, so I will merely claim that reductionism takes nothing from the whole; it adds new layers of wonder to the experience. That applies whether the designer of the parts was a human being or the GOD.

So how does a gosling's brain imprint on a professor? Until very recently it was a complete mystery. Within the past few years, though, the veils of mystery have begun to lift, revealing new veils beneath. The first veil concerns which part of the brain is involved. When a chick imprints on its parents, experiments reveal that memories are laid down first and most rapidly in a part of the brain called the left IMHV (intermediate and medial hyperstriatum ventrale). In this part of the brain, and only on the left side, a whole rash of changes accompanies imprinting: neurons change shape, synapses form and genes are switched on. If the left IMHV is damaged, the chick fails to imprint on its mother.

The second veil to lift reveals which chemical is necessary for 'filial' imprinting of this kind. By examining the brains of chicks after they had or had not imprinted on an object, Brian McCabe found that a neurotransmitter called GABA is released from brain cells in the left IMHV during imprinting. He had previously noticed that a gene for a GABA receptor is switched off about ten hours after the chick has been trained to imprint on an object.[21]

So something happens in one part of the left side of the chick's brain during imprinting first to release GABA and then to reduce sensitivity to GABA at the end of the critical period. To take the story further, it is time to leave baby birds for a different kind of critical period, one that is a little easier to study: the development of binocular vision. Babies are occasionally born with cataracts in both eyes that render them blind. Until the 1930s surgeons thought it wise not to operate to remove such cataracts until after the age of ten, because of the risks of surgery on small children. But it became apparent that such children never managed to perceive depth or shape properly even after the removal of the cataracts. It was simply too late for the visual system to learn 'how to see'. Likewise, monkeys reared in darkness for the first six months of their lives took months to learn to distinguish circles from squares, something normal monkeys could learn in days. Without visual experience in the first months of life, the brain cannot interpret what the eye sees. A critical period has passed.

There is one layer of primary visual cortex, called layer 4C, that receives inputs from both eyes and separates them into streams from each eye. To begin with, the inputs are randomly distributed, but before birth they become roughly sorted into stripes, each stripe responding mainly to one eye. During the first few months of life after birth, this segregation becomes increasingly marked, so that all the cells responding to the right eye become clustered into right-eye stripes, while all those responding to the left eye become clustered into left-eye stripes. These stripes are called ocular dominance columns. Amazingly, the columns do not segregate in the brains of animals deprived of sight during the early months of life.

David Hubel and Torsten Wiesel discovered how to stain these columns different colours by injecting dyed amino acids into one eye. They were then able to see what happens when one eye is sewn shut. In an adult animal, this has virtually no effect on the stripes. But if one eye is sewn shut for as little as a week during the first six months of a monkey's life, then the stripes from that deprived eye almost disappear and that eye becomes effectively blind, because it has nowhere in the brain to which to report. The effect is irreversible. It is as if the neurons from the two eyes compete for space in layer 4C and those that are active win the battle.

These experiments in the 1960s were the first demonstrations of 'plasticity' in the development of the brain during a critical period after birth. That is to say, the brain is open to calibration by experience in the early weeks of life, after which it sets. Only by experiencing the world through its eyes can an animal sort the input into separate stripes. Experience seems actually to switch on certain genes, which in turn switch on others.[22]

By the late 1990s, a number of people were searching for the molecular key to this critical period of plasticity in vision. Their method of choice was genetic engineering: the creation of mice with extra genes or missing genes. Mice, like cats and monkeys, have a critical period during which the inputs from the two eyes compete for space in the brain, though they do not sort into neat columns. In Boston, in the lab of Susumu Tonegawa, Josh Huang reckoned he had an idea of what they were competing for: brain-derived neurotrophic factor, or BDNF, the product of a gene one version of which also seems to predict neurotic personalities (see chapter 3). BDNF is a sort of brain food: it encourages the growth of neurons. Perhaps, reasoned Huang, the cells carrying the most signals from the eye got more BDNF than the silent cells, so the input from the open eye displaced the input from the closed eye. In a world where there was not enough BDNF to go round, it was survival of the hungriest neuron.

Huang did the obvious experiment: he made a mouse that produced extra BDNF from its genes, expecting that this BDNF would now provide ample food for all neurons, enabling the input from both eyes to survive. He was surprised to see a different and dramatic effect. The mice with extra BDNF went through the critical period faster. Their brains set two weeks after eye opening instead of three. This was the first demonstration that a critical period could be adjusted artificially.[23]

A year later in 2000 came another breakthrough in the laboratory of a Japanese scientist, Takao Hensch. Hensch discovered that a mouse lacking a gene called GAD65 failed to sort its eye inputs in response to visual stimuli. But these same knock-out mice did sort their inputs if injected with the drug diazepam. Indeed, diazepam, like BDNF, seemed to bring on a precocious imprinting. Injecting diazepam after the critical period could not restore plasticity to the brain. In the GAD65 knock-out mice, the scientists could bring on plasticity with diazepam at any time, even during adulthood. But only once. After the reorganisation caused by diazepam, the system entirely lost its sensitivity. It is as if there is a dormant program for rewiring the brain, which can be triggered once – but only once.[24]

Back in Boston Huang had surprised himself again. Together with Lamberto Maffei in Pisa, he had simply reared his transgenic mice – the ones with the extra BDNF – in the dark. Normal mice, if raised in the dark for three weeks after eye opening, are effectively blind for life; they need the experience of light to mature their visual system. To put it bluntly their brains need nurture, as well as nature. But remarkably, the extra-BDNF mice reared in the dark responded normally to visual stimuli, suggesting that they could see well despite having had no exposure to light during the critical period. Huang and Maffei had stumbled on an extraordinary fact: a gene that could substitute for aspects of experience. One of the roles of experience is apparently not to fine-tune the brain, but merely to switch on the BDNF gene, which in turn fine-tunes the brain. If you shut the eye of a mouse, BDNF production in its visual cortex drops within half an hour.[25]

Despite this result, Huang does not really believe experience is dispensable. He notes that the system seems to be designed to delay maturation of the brain until experience is available. What do BDNF, GAD65 and diazepam – the three things that can affect critical periods – have in common? The answer is the neurotransmitter GABA: GAD65 makes it, diazepam mimics it and BDNF regulates it. Since GABA was implicated in filial imprinting in the chick, it looks plausible that the GABA system will prove to be central to critical periods of all kinds. GABA is a sort of neuronal spoil-sport: it inhibits the firing of neighbouring neurons. Feeling unloved, the inhibited neurons die back. Because the maturation of the GABA system is itself dependent on visual experience and is BDNF-driven, the link between them has the ring of truth.

Though it is still far from complete, the GABA story is a beautiful example of how it is now possible as it never was before to begin to understand the molecular mechanisms behind such things as imprinting. It shows just how unfair is the charge that reductionism takes the poetry out of life. Who would have conceived of a mechanism so exquisitely designed if they had refused to look under the lid of the brain? Only by equipping the brain with BDNF and GAD65 genes, can the GOD make a brain capable of absorbing the experience of seeing. These are, if you like, the genes for nurture.

YOUNG TONGUES

Critical-period imprinting is everywhere. There are a thousand ways in which human beings are malleable in their youth, but fixed once adult. Just as a gosling is imprinted with an image of its mother during the hours after birth, so a child is imprinted with everything from the number of sweat glands on its body and a preference for certain foods to an appreciation of the rituals and patterns of its own culture. Neither the gosling's mother-image nor the child's culture is in any sense innate. But the ability to absorb them is.

An obvious example is accent. People change their accents easily during youth, generally adopting the accent of people of their own own age in the surrounding society. But some time between about 15 and 25, this flexibility simply vanishes. From then on, even if somebody emigrates to a different country and lives there for many years, his accent will change very little. He may pick up a few inflections and habits from his new linguistic surroundings, but not many. This is true of regional as well as national accents: adults retain the accent of their youths; youths adopt the accent of the surrounding society. Take Henry Kissinger and his younger brother Walter. Henry was born on 27 May 1923, while Walter was born just over a year later on 21 June 1924. They both moved to the United States as refugees from Germany in 1938. Today Walter sounds like an American, whereas Henry has a characteristic European accent. A reporter once asked Walter why Henry had a German accent but he did not. 'Because Henry doesn't listen,' came the facetious reply. It seems more likely that when they arrived in American Henry was just old enough to be losing the flexibility of imprinting his accent on his surroundings; he was leaving the critical period.

In 1967 a Harvard psychologist, Eric Lenneberg, published a book in which he argued that the ability to learn language is itself subject to a critical period that ends abruptly at puberty. Evidence for Lenneberg's theory now abounds on all sides, not least in the phenomenon of creole and pidgin languages. Pidgins are languages used by adults of several different linguistic backgrounds to communicate with each other. They lack consistent or sophisticated grammar. But once they have been learnt by a generation of children still in their critical period, they change into creoles – new languages with full grammar. In one case in Nicaragua, deaf children sent to new deaf schools together for the first time in 1979 simply invented a new sign-language creole of remarkable sophistication.[26]

But the most direct test of the critical period in language learning would be to deprive a child of all language until the age of 13 and then try to teach the poor creature to speak. Deliberate experiments of this kind are thankfully rare, though at least three monarchs – King

Psamtik of Egypt in the seventh century BC, the Holy Roman Emperor Frederick II in the thirteenth century and King James IV of Scotland in the fifteenth century – are said to have tried depriving newborn children of all human contact except a silent foster-mother to see whether they grew up speaking Hebrew, Arabic, Latin or Greek. In Frederick's case, the children all died. In a variant on the practice, the Moghul emperor Akbar is said to have done the same experiment to find out whether people were innately Hindu, Muslim or Christian. All he got was deaf-mutes. Genetic determinists were made of stern stuff in those days.

By the nineteenth century, attention had shifted to natural deprivation experiments in the form of 'feral children'. Two seem to have been genuine. The first was Victor, the wild boy of Aveyron, who appeared in 1800 in the Languedoc having apparently lived wild for many of his 12 years. Despite years of effort, his teacher failed to teach him to speak and 'abandoned my pupil to incurable dumbness'.[27] The second was Kaspar Hauser, a young man discovered in Nuremberg in 1828 having apparently been kept in a single room with almost no human contact for all of his 16 years. Even after years of careful coaching, Kaspar's syntax was still 'in a state of miserable confusion'.[28]

Two suggestive cases, but hardly proof. Then suddenly, four years after Lenneberg's book, there was a third case of a wild child first found after puberty: a 13-year-old girl named Genie was discovered in Los Angeles after a childhood of almost inconceivable horror. The daughter of a blind, abused mother and a paranoid and increasingly reclusive father, she had been kept in silence in a single room, mostly either harnessed to a potty chair or confined in a caged cot. She was incontinent, deformed and almost completely mute: her vocabulary consisted of two words: 'Stopit' and 'Nomore'.

The story of Genie's rehabilitation is almost as tragic as that of her childhood. As she was passed between scientists, foster-parents, state officials and her mother (the father committed suicide after her discovery), the initial optimism of those who set out to care for her was gradually spent in lawsuits and bitterness. Today she is in a home for retarded adults. She learned much, her intelligence was high, her non-verbal communication was extraordinary and her ability to solve spatial puzzles was ahead of her age.

But she never learned to speak. She developed a good vocabulary, but elementary grammar was beyond her, and the syntax of word order was a foreign land. She could not grasp how to phrase a question by altering word order or how to change 'you' to 'I' in an answer. (Kaspar Hauser had the same problem.) Though the psychologists who studied her at first believed she would disprove Lenneberg's critical-period theory, they eventually admitted that she was a confirmation of it. Untrained by conversation, the brain's language module had simply not developed, and it was now too late.[29]

Victor, Kaspar and Genie (and there have been other cases, including a woman not diagnosed as deaf until she was 30) suggest that language does not just develop according to a genetic programme. Nor is it just absorbed from the outside world. Instead it is imprinted. It is a temporary innate ability to learn by experience from the environment, a natural instinct for acquiring nurture. Polarise that into either nature or nurture, if you can.

Though language was the most severe of Genie's problems in adjusting to the world, it was not the only one. After her release she became an obsessive collector of coloured plastic objects. She was also for many years terrified of dogs. Both of these characteristics could be tentatively traced to 'formative experiences' in her childhood. Just about the only toys she had were two plastic raincoats. As for dogs, her father would bark and growl outside her door to frighten her if she made a noise. How many of a person's own preferences, fears and habits are imprinted during her youth? Most of us can recall in astonishing detail the places and people of our early years, whereas we forget much more recent adult experiences. Memory is plainly not all critical period – it does not switch off at a certain age. But there is an element of truth in the old notion that the child is father to the man. Freud was right to emphasise the importance of formative years, even if he sometimes generalised too freely about them.

FAMILIARITY BREEDS INDIFFERENCE

One of the more controversial theories of human imprinting concerns incest. The critical period in the development of sexual orientation plainly leaves a young person committed to being attracted to members of the opposite sex (except when it makes them committed to being attracted to members of the same sex). Probably it also determines 'your type' of partner in some much more specific way. But does it also determine who you will be positively averse to wooing?

The law forbids brother–sister marriage and for good reason. Inbreeding causes horrific genetic diseases by bringing together rare recessive genes. But suppose some country were to repeal that law and proclaim that from now on, brother–sister marriages were not only legal, but rather a good thing. What would happen? Nothing. Despite being the best of friends and highly compatible, most women are simply not attracted to their brothers 'in that way'. In 1891, a Finnish pioneer of sociology named Edward Westermarck published a book called *The History of Human Marriage* in which he suggested that human beings avoid incest by instinct rather than by obedience to the law. They are naturally averse to sex with close kin. Cleverly, he saw that this did not require people to have an innate ability to recognise real brothers and sisters. Instead, there was a rough-and-ready way of knowing: those people whom one had known well as children were probably close kin. He predicted that people who have shared childhood will be instinctively averse to sleeping with one another as adults.

Within 20 years Westermarck's idea was all but forgotten. Freud criticised his theory and suggested instead that human beings were attracted to incest and were only prevented from practising it by cultural prohibitions in the form of taboos. Oedipus without incestuous desire is like Hamlet without madness. But if people are averse to incest, they cannot have incestuous desires. And if they need taboos, it means they must have desires. Westermarck protested in vain that social learning theories 'imply that the home is kept free from incestuous intercourse by law, custom, or education. But even if social prohibitions might prevent unions between the nearest relatives, they could not prevent the desire for such unions. The sexual instinct is hardly changed by proscriptions.'[30]

Westermarck died in 1939 as Freud's star was still rising and 'biological' explanations were falling out of fashion. It took another 40 years before somebody looked again at the facts. That somebody was a sinologist named Arthur Wolf, who analysed the meticulous demographic records kept by the occupying Japanese in nineteenth-century Taiwan. Wolf noticed that these long-dead Chinese had practised two forms of arranged marriage. In one, the bride and groom met on their wedding day, though the match was arranged many years before. In the other, the bride was adopted by the groom's family as an infant and reared by her future in-laws. Wolf realised that this was a perfect test of the Westermarck hypothesis, for these 'sim-puahs' or 'little daughter-in-laws' would experience the illusion that they were expected to marry their brother. If, as Westermarck argued, shared childhood led to sexual aversion, then these marriages ought not to work very well.

Wolf collected information on 14,000 Chinese women and compared those who had been sim-puahs with those who met their arranged husbands only on their wedding day. Astonishingly, marriage to a childhood associate was 2.65 times as likely to end in divorce as an arranged marriage to an unfamiliar partner – people who had known each other all their lives were much less likely to stay married than people who had never met! The sim-puah marriages also produced fewer children and experienced more adultery. Wolf ruled out other obvious explanations – that the process of adoption led to ill health and infertility, for example. Far from bringing spouses together, the habit of co-rearing them seemed to inhibit the later development of sexual attraction. But this was only true of sim-puahs adopted at the age of three or younger; those adopted at four or older had just as successful marriages as those who met as adults.[31]

Since then many studies have confirmed the same phenomenon. Israelis reared communally in a kibbutz rarely marry each other.[32] Moroccans who have slept in the same room as children are averse to accepting arranged marriage.[33] The aversion seems to be stronger among women than men. Even in fiction, echoes of the aversion reverberate: Victor Frankenstein, in the Mary Shelley novel, finds himself expected to marry a cousin reared with him since childhood – but (symbolically) his monster intervenes to kill his prospective bride before the marriage is consummated.[34]

It is true that incest taboos exist, but on closer inspection they are little concerned with close-kin marriage. They are all about regulating the marriages between cousins.[35] It is true, also, that people seem to be fascinated by incest, and that it plays a large part in medieval fiction, Victorian scandal and modern urban legends. But then things that horrify people also fascinate them: snakes often fascinate as much as they horrify. It also seems to be true that siblings separated at birth who later find each other as adults are often strongly attracted to each other,[36] but this if anything supports the Westermarck effect.

The Westermarck effect is plainly not universally effective. Exceptions do exist both at the cultural and at the individual level. Many sim-puah brides were able to overcome their sexual aversion and have successful marriages: the system had set their incest-avoidance instinct against an even stronger instinct for procreation. Also there is some evidence that 'fooling around' between brothers and sisters who were reared together does occur, whereas those who had been separated for more than a year during early childhood were much more likely to have indulged in actual intercourse. In other words, childhood association may not produce an aversion to attraction, so much as to actual intercourse.[37]

None the less, incest-aversion between those reared in the same family, like language, seems to be a clear case of a habit imprinted on the mind during a critical period of youth. In one sense it is pure nurture – the mind has no preconceptions about whom it will become averse to, so long as they are childhood companions. And yet it is nature in the sense of an inevitable development set in train presumably by some genetic programme at a particular age. Author's message: you need nature to be able to absorb nurture.

Just like Lorenz's goslings in reverse, we are imprinted with an aversion rather than an attachment. So here's a funny thing: Konrad Lorenz married his childhood friend Gretl, the girl with whom he imprinted his first duckling at the age of six. She was the daughter of a market gardener in the next-door village. Why were they not averse to each other? Perhaps a clue lies in the fact that she was three years older than him. This means she was probably already out of the critical period for the Westermarck effect by the time they grew to know each other. Or perhaps Konrad Lorenz was just an exception to his own rule. Biology, somebody once said, is the science of exceptions, not rules.

NAZITOPIA

Lorenz's notion of imprinting was a great insight that has stood the test of time. It is a crucial part of the nature-via-nurture jigsaw and an exquisite marriage of the two. The invention of imprinting as a way of ensuring the flexible calibration of instinct was one of natural selection's master strokes. Without it we would all either be born with a fixed and inflexible language unchanged since the Stone Age, or we would struggle to relearn each grammatical construction. But one of Lorenz's other ideas will not be so kindly judged by history. Though the story has little to do with imprinting, it is worthwhile to recount how Lorenz, like so many others, fell into the twentieth century's usual trap of flirting with a sort of utopia.

In 1937 Lorenz was unemployed. His studies of animal instinct were prohibited in the Catholic-dominated University of Vienna on theological grounds and he had retired to Altenberg to continue his bird work at his own expense. He applied for a grant to work in Germany. Commenting on the application, a Nazi official wrote: 'All reviews from Austria agree that the political attitude of Dr Lorenz is impeccable in every respect. He is not politically active, but in Austria he never made a secret of the fact that he approved of National Socialism . . . Everything is also in order with his Aryan descent.' In June 1938, shortly after the *Anschluss*, Lorenz joined the Nazi Party and became a member of the party's Office of Race Policy.

He immediately began speaking and writing about how his work on animal behaviour could fit in with Nazi ideology; in 1940 he was appointed a professor at the University of Königsberg. Over the next few years, until his capture on the Russian front in 1944, he argued consistently in favour of the utopian ideals of 'a scientifically underpinned race policy', 'the racial improvement of Volk and race', and the 'elimination of the ethically inferior'.

After suffering in a Russian prisoner-of-war camp for four years at the end of the war, Lorenz returned to Austria. He managed to gloss over his Nazi actions as gullible and stupid, but said he was not politically active. It was more that he tried to bend his science to suit the new political powers than that he genuinely believed in it, he said. While he lived, this was accepted. But after he died it gradually emerged how deeply he had imbibed Nazism. In 1942, while serving as a military psychologist in Poland, Lorenz took part in research led by psychologist Rudolf Hippius and sponsored by the SS, the aim of which was to develop criteria for distinguishing 'German' from 'Polish' features of 'half-breeds' in order to help the SS decide which to choose for their 'Re-Germanisation' effort. There is no evidence that he was involved in war crimes himself, but he probably knew that they were committed.[38]

Central to his argument, during this Nazi period, was the issue of domestication. Lorenz had developed a rather quaint contempt for domesticated animals, which he regarded as greedy, stupid and oversexed compared with their wild relatives. 'Great ugly beast,' he once cried while rejecting the sexual advances of an imprinted muscovy duck.[39] Pejoratives aside, he had a point. Almost by definition, selective breeding for domesticity produces animals that fatten well, breed well and are docile and dull. Cows and pigs have brains that are one-third smaller than their wild relatives. Female dogs are fertile twice as often as wolves. And pigs notoriously can gain far more weight than wild boar.

Lorenz began to apply these notions to humanity. In a notorious 1940 paper, entitled 'Disorders caused by the domestication of species-specific behaviour', he argued that human beings are self-domesticated and that this has led them into physical, moral and genetic deterioration. 'Our species-specific sensitivity to the beauty and ugliness of members of our species is intimately connected with the symptoms of degeneration caused by domestication, which threatens our race . . . The racial idea as the basis of our state has already accomplished much in this respect.' In effect, Lorenz's domestication argument opened a new front in the eugenic debate, giving another reason to nationalise reproduction and eliminate both unfit individuals and unfit races. Lorenz seems not to have spotted a large flaw in his own argument, that the muscovy duck is inbred after generations of selection to narrow its gene pool, whereas civilisation has the opposite effect on people: it relaxes selection, allowing more mutations to survive in the gene pool.

There is no evidence that this had any influence on Nazism, which already had plenty of reasons, some more 'scientific' than others, for its policies of racism and genocide. Lorenz's argument was ignored, perhaps even distrusted, by the party. What is more remarkable, perhaps, is that Lorenz's domestication argument survived the war, to be reiterated in less emotive terms in his book *Civilized Man's Eight Deadly Sins*, first published in 1973. This was a book that combined Lorenz's older concerns about human degeneration caused by the relaxation of natural selection with newer and more fashionable concerns about the state of the environment. As well as genetic deterioration, the eight deadly sins were overpopulation, destruction of the environment, over-competition, the seeking of instant gratification, indoctrination by behaviourist techniques, the generation gap and nuclear annihilation.

Genocide was not on Lorenz's list.

CHAPTER SEVEN

Learning lessons

'All men are similar, in soul as well as body. Each of us has a brain, spleen, heart and lungs of similar construction; and the so-called moral qualities are the same in all of us – the slight variations are of no importance . . . Moral diseases are caused by the wrong sort of education, by all the rubbish people's heads are stuffed with from childhood onwards, in short by the disordered state of society. Reform society and there will be no diseases . . . At any rate, in a properly organised society it won't matter a jot whether a man is stupid or clever, bad or good.'

'Yes, I see. They will have identical spleens.'

'Precisely, madame.'

Bazarov and Madame Odintsov, in *Fathers and Sons*, by Ivan Turgenev.[1]

In 1893 Alfred Nobel, the Swedish inventor of dynamite, was beginning to feel his age. Over 60 and not in good health, he heard rumours that miraculous feats of rejuvenation might be achieved with transfusions of blood from giraffes. When rich men are in this kind of mood, the astute scientist gets out the begging bowl. Nobel was duly persuaded to pay 10,000 roubles for a grand new physiology building for Russia's Imperial Institute of Experimental Medicine outside

St Petersburg. Nobel died anyway in 1896 and the laboratory never bought a giraffe, but it went from strength to strength. With a staff of over 100, and managed like a business, it was a sort of scientific factory. In charge was an ambitious and confident young man named Ivan Petrovich Pavlov.[2]

Pavlov was a disciple of Ivan Mikhailovich Sechenov, who was so obsessed with reflexes that he believed that thought was nothing but a reflex with the action missing. He was as dedicated to the cause of nurture as his contemporary Galton was to the cause of nature: he believed that 'the real cause of every activity lies outside man' and that '999/1,000 of the contents of the mind depends on education in the broadest sense, and only 1/1,000 depends on individuality'.[3]

Sechenov's philosophy guided much of the torrent of experimental work that poured from Pavlov's factory over the next three decades. The victims of these experiments were mostly dogs, or 'dog technologies' as they were rather coldly called. At first Pavlov concentrated on the digestive glands of the dog; later he began to move into the brain. In 1903 at a conference in Madrid, he announced the results of his most famous experiment. It had begun, like so much great science, serendipitously. He was trying to study the dog's salivation reflex in response to food, and had diverted one of a dog's salivary glands into a funnel so he could measure the production of saliva. The dog, however, would start salivating as soon as it heard the food being prepared, or even as soon as it was strapped into the apparatus – anticipating the food.

This 'psychic reflex' was not what Pavlov was after, but he suddenly saw its significance and switched his attention to it. The dog was now led to expect food whenever it heard a bell or a metronome, and it soon began to salivate to the sound of the bell alone. Pavlov having diverted its salivary glands into a funnel, he could actually count the drops of saliva produced in response to each ring of the bell. Later he proved that a dog with no cerebral cortex could still reflexively salivate when fed, but not when alerted by the bell. The 'conditioned reflex' to the bell therefore lay in the cortex itself.[4]

Pavlov seemed to have discovered a mechanism – conditioning, or association – by which the brain could acquire its knowledge of the regularities of the world. It was a great discovery, it was right and of course it was not the whole answer. But as usual, some of Pavlov's followers went too far. They began to assert that the brain was nothing but a device for learning through conditioning. This tradition flowered in the United States as behaviourism. Its champion was John Broadus Watson, of whom more later.

Modern learning theorists have modified Pavlov's idea in one crucial way. They argue that the active learning occurs not when the stimulus and reward continue to appear together, but when there is some discrepancy between an expected coincidence and what actually happens. If the mind makes a 'prediction error' – expecting a reward after a stimulus and not getting it, or vice versa – then the mind must change its expectation: it must learn. So, for example, if the bell no longer predicts the food, but a flash of light now does predict the food, the dog must learn from the discrepancy between its own expectations and the new reality. Surprise, pleasant or unpleasant, is more informative than predictability.

This new emphasis on prediction errors now takes physical form in the brain as well as psychological form in the mind. In a series of experiments on monkeys, Wolfram Schultz has discovered that dopamine-secreting neurons in a certain part of the brain (the substantia nigra and ventral tegmental area) react to surprise, but not to predicted effects. They fire more when the monkey is rewarded and less when it is unexpectedly deprived of a reward. The dopamine cells themselves, in other words, actually encode the same rule of learning theory that engineers now try to build into robots.[5]

Pavlov, the indefatigable dissector of dogs, would have enjoyed such a reductionist result. But he might have been made uneasy by a philosophical irony this result leads to. He was out to prove that the dog's brain learned about its situation from the world, that in Sechenov's words 'the real cause . . . lies outside man'. He stood in a long tradition of empiricism stretching back through Mill and Hume to Locke: human nature was largely the scribbling of experience on the blank sheet of the mind. Yet for the mind to scribble on its sheet, it must have dopamine neurons specially designed to respond to surprise. And how are they so designed? By genes. Today precisely the equivalent experiment that Pavlov performed is being done, routinely, in many of the top genetics laboratories of the world, because Pavlov's modern descendants are busy proving the role that genes play in learning. Here lies the proof of this book's theme: genes are not only involved in nature; they are just as intimately involved in nurture, too.

The modern Pavlovian experiment is often done with fruit flies, but the principle is identical. A fly is given an electric shock through its feet shortly after a puff of smelly chemical is squirted into its test tube. Pretty soon the fly learns that the smell will be followed by the shock, so it takes to the air before the shock arrives: it has made the (initially surprising) association between the two phenomena. This experiment was first done by Chip Quinn and Seymour Benzer in the 1970s at the California Institute of Technology. It proved, to universal surprise, that flies can learn and remember associations between smells and shocks.

It also proved that they can only do so if they have certain genes. Mutants missing a crucial gene just don't get the point. There are at least 17 genes that are essential to the laying down of a new memory in the fruit fly. These genes have pejorative names – dunce, amnesiac, cabbage, rutabaga and so on – which is a bit unfair, since the fly is only a dunce if it lacks the gene, not if it has it. Recognisably the same set of so-called CREB genes is used by all animals including human beings. They must be turned on – that is, they must create a protein – during the learning process itself.

This is an astonishing discovery, rarely appreciated for quite how shocking it is. Here is what John B. Watson said about associative learning in 1914:

> Most of the psychologists talk quite volubly about the formation of new pathways in the brain, as though there were a group of tiny servants of Vulcan there who run through the nervous system with hammer and chisel digging new trenches and deepening old ones.[6]

Watson was mocking the idea. But the joke is on him. The formation of a mental association takes the form of new and strengthened connections between neurons. The servants of Vulcan that create those connections exist. They are called genes. Genes! Those implacable puppet masters of fate that are supposed to make the brain and leave it to get on with the job. But they do not; they also actually do the learning. Right now, somewhere in your head, a gene is switching on, so that a series of proteins can go to work altering the synapses between brain cells so that you will, perhaps, for ever associate reading this paragraph with the smell of coffee seeping in from the kitchen . . .

I cannot emphasise the next sentence strongly enough. These genes are at the mercy of our behaviour; not the other way round. The things that make Pavlov's associations are made of the same stuff as the chromosomes that carry heredity. Memory is 'in the genes' in the sense that it uses genes, not in the sense that you inherit memories. Nurture is effected by genes just as much as nature is.

Here follows one example of such a gene. In 2001, Josh Dubnau working with Tim Tully did an exquisite experiment on a fruit fly. Please wallow in the details of the methods for a few moments just to appreciate the sophistication of the tools available to modern molecular biology (and then pause to reflect just how much more sophisticated they will be in a few years' time). First, he made a temperature-sensitive mutation in a particular fly gene, called shibire, the gene for a motor protein called dynamin. This means that at 30 degrees C the fly is paralysed, but at 20 degrees C it recovers completely. Next he engineered a fly in which this mutant gene is active only in the output from one part of the fly's brain, called the mushroom body, which is essential for learning to associate smells with shocks. This fly is not paralysed at 30 degrees C, but it cannot retrieve memories. When such a fly is trained, while hot, to pair a smell with danger, then asked, when cool, to retrieve the memory, it performs well. In the opposite circumstance, when the fly is asked to form the memory while cool and retrieve the memory while hot, it cannot.[7]

Conclusion: the acquisition of a memory is distinct from its retrieval; different genes are needed in different parts of the brain. The output from the mushroom body is necessary for retrieval, but not for acquisition of memory, and the switching on of a gene is necessary for that output. Pavlov may have dreamed that one day somebody would understand the wiring in the brain that explained associative learning, but he surely could not have imagined that somebody would go still deeper and describe the actual molecules, let alone find that the key to the process, minute by minute, lies with Gregor Mendel's little particles of heredity.

This is a science in its infancy. Those who study the genes involved in learning and memory have struck a rich seam of ignorance to mine. Tully, for instance, has now set himself the immense task of understanding how these genes of memory alter some of the synapses between their home neuron and its neighbour while leaving other synapses untouched. Each neuron has on average 70 synapses connecting it to other cells. Somehow, in the cell nucleus, the CREB gene on chromosome 1 has the job of switching on a set of other genes and those other genes must then send their transcripts to just the right synapses where they can be used to change the strength of the connection. Tully has at last found a way to understand how that is done.[8]

Yet CREB is only part of the story. Seth Grant has found evidence that many of the genes necessary for learning and memory are not so much part of a sequential network; more, they go to make up a machine, which he calls a Hebbosome (for reasons that will become clear later). One such Hebbosome consists of at least 75 different proteins – that is, the products of 75 genes – and appears to work as a single complex machine.[9]

MAKING BABIES CRY

I promised to return to John B. Watson. Reared in rural poverty and isolation in South Carolina, Watson was the son of a devout mother and a philandering father who left home when he was 13. This background gave him – either through genes or experience – a strong and truculent character. He was a violent adolescent, a faithless husband and a domineering father, who drove a son to suicide and a granddaughter to drink, before becoming a bitter recluse in retirement. He also caused a revolution in the study of human behaviour. Frustrated by the waffle that passed for psychology, in 1913 he outlined a bold manifesto for reform in a lecture entitled 'Psychology as the behaviorist views it'.[10]

Introspection, he announced, must cease. Legend has it that Watson was disgusted to be asked to imagine what went on in the mind of a rat as it ran through a maze. He suffered from physics envy. The science of psychology must be put on an objective foundation. Behaviour, not thought, was what counted. 'The subject matter of human psychology is the behavior of the human being.' In other words, the psychologist should study what went into the organism and what came out, not the processes in between. The principles that governed learning could be derived from any animal and applied to people.

Watson drew his ideas from three main streams of thought. William James, though himself a nativist, had stressed the role of habit formation in human behaviour. Edward Thorndike had gone further, coining his 'Law of Effect' whereby animals repeated actions that produced pleasant results and did not repeat actions that had unpleasant consequences: an idea that also goes under the names reinforcement learning, trial-and-error learning, instrumental conditioning or operant conditioning (these psychologists love their jargon). In Thorndike's experiments, a cat had found the lever to open the door to its cage by trial and error; within a few trials it knew exactly how to open the door. Though Pavlov's work was not translated until 1927, Watson knew of it from his friend Robert Yerkes, and saw immediately that Pavlovian or classical conditioning was a centrepiece of learning. At last here was a psychologist as rigorous as physicists: 'I saw the enormous contribution Pavlov had made, and how easily the conditioned response could be looked upon as the unit of what we had all been calling HABIT.'[11]

In 1920, Watson and his assistant Rosalie Rayner performed an experiment that convinced him that emotional reactions could be conditioned, and that human beings could be treated as large, hairless rats. It was an immensely influential experiment. A word about Rayner is relevant here. She was the 19-year-old niece of a prominent senator famous for conducting hearings into the sinking of the *Titanic*. She was beautiful and rich, and she drove about Baltimore in a Stutz Bearcat. Watson fell in love with her and she with him. Watson's wife found a love letter from Rayner in his coat, but she was advised by a lawyer to see if she could find a letter from him, not to him, before confronting him. So she went round to the Rayners' house for coffee, where she feigned a headache and asked to lie down. Upstairs, she quickly locked herself in Rosalie's bedroom and searched it, finding 14 love letters from her husband. The ensuing scandal cost Watson his academic career. He divorced his wife, married Rayner and left psychology for an advertising career with J. Walter Thompson, where he devised a successful campaign for Johnson's baby powder and persuaded the Queen of Romania to endorse Pond's facial cream.

The subject of the 1920 experiment between these lovebirds was a little child called Albert B, who had been reared from birth in a hospital. (It has been claimed that Albert was Watson's illegitimate child by a nurse, but I can find no proof of this.) When Albert was 11 months of age, Watson and Rayner showed him a series of objects, including a white rat. None of the objects frightened Albert; he enjoyed playing with the rat. But when they suddenly banged a hammer on a steel bar, Albert cried, not unreasonably. The two psychologists then began banging the bar whenever Albert touched the rat. Within a few days Albert was liable to start crying as soon as the rat appeared, a conditioned fear response. He was now frightened of a white rabbit, too, and even a sealskin coat, apparently having transferred his fear to any white, furry thing. With characteristic sarcasm, Watson announced the moral of the tale:

> The Freudians, twenty years from now, unless their hypotheses change, when they come to analyse Albert's fear of a sealskin coat – assuming he comes to analysis at that age – will probably tease from him the recital of a dream which upon their analysis will show that Albert at three years of age

attempted to play with the pubic hair of the mother and was scolded violently for it.[12]

(If you ask me it is Watson who needs scolding.)

By the mid-1920s Watson was convinced not that conditioning was a part of how humans learned about the world, but that it was the main theme. He joined a growing academic enthusiasm for nurture over nature and made the extraordinary claim:

Give me a dozen healthy infants, well-formed, and my own specified world to bring them up in and I'll guarantee to take any one of them at random and train him to become any type of specialist I might select – doctor, lawyer, artist, merchant-chief, and yes, even beggar-man and thief, regardless of his talents, penchants, tendencies, abilities, vocations and race of his ancestors.[13]

REDESIGNING PEOPLE

Ironically, five years before Watson's claim, a very powerful man had had the same thought: Vladimir Ilyich Lenin. Lenin, like Pavlov, was influenced by the environmentalism of Sechenov, via the writings of Nikolai Chernyshevsky. Two years after the Russian Revolution, Lenin is said to have paid a secret visit to Pavlov's physiology factory and asked him if it was possible to engineer human nature.[14] No record of the meeting survives, so Pavlov's views on the matter are unknown. Perhaps he had more pressing concerns: with the famine induced by the civil war, the institute's dogs were starving, and the researchers could keep them alive only by sharing their meagre rations with them. Pavlov had begun to cultivate his own vegetable patch at the institute, leading by example and driving his students to feats of horticulture as energetically as he had driven them to feats of science.[15] No hint of political encouragement to Lenin from Pavlov comes down to us. Pavlov was an outspoken critic of the revolution, though he mellowed when shown favour by the commissars.

Lenin could undoubtedly see that the success of communism rested on an assumption that human nature could be trained to a new system. 'Man can be corrected,' he said. 'Man can be made what we want him to be.' Echoed Trotsky: 'To produce a new, "improved version" of man – that is the future task of Communism.'[16] Much Marxist debate revolved around the question of how long it would take to produce a 'new man'. Such an aim makes no sense unless human nature is almost entirely malleable. In this sense, communism always had a vested interest in nurture rather than nature. But the state was slow to put this idea into practice. In the 1920s, even the Soviet Union caught the global enthusiasm for eugenics. N.A. Semashko outlined an ambitious programme of socialist eugenics in 1922, celebrating the appalling truth that eugenics 'will place the interests of the whole society, of the collective, first, above the interests of the individual persons'. The 'new man' was to be bred. But under Stalin, Soviet eugenics collapsed, as communist leaders realised that not only would this take several generations, but preserving the intelligentsia by selective breeding rather contradicted the general secretary's increasingly obvious preference for persecuting intellectuals. After the Nazis came to power in Germany, there was another reason to reject eugenics: the study of human heredity was equated with the rival creed of fascism. Russian eugenicists were soon criticised for their hereditarian beliefs; for not 'grasping the social levers' instead.[17]

The person who would grasp the social levers came from an unexpected direction. In the 1920s, with Russia in the grip of famine, the government discovered an elderly and paranoid crank who bred apples near Kozlov, called Ivan Vladimirovich Michurin. Michurin made absurd claims – that he could make a pear sweeter to the second generation by watering it with sugar water, or that grafting produced a hybrid stock. He suddenly found himself showered with honours and grants by a government desperate for quick ways of boosting food production. Michurinism was promoted as a new science to replace Mendelism.

The scene was set for a scientific coup. A young man called Trofim Denisovich Lysenko managed to catch the attention of *Pravda* for his ability to breed a better crop of wheat by Michurinist means. At the time, winter-sown wheat was killed by winter frost except in the far south of the country, while spring-sown wheat sometimes came into ear too late and was killed by drought. Lysenko at first claimed to have bred hardy winter wheat by 'training' it. By 1928–9, seven million hectares were planted with his technique: it all died. Unfazed, Lysenko switched to spring wheat, claiming that simple soaking – vernalisation – would render it quick to ear. Again this merely exacerbated the famine. By 1933 vernalisation had been dropped.

But Lysenko, who was better at politics than science, went from strength to strength and was soon touting his ideas as a new form of science that disproved the theory of the gene and demolished the tenets of Darwinism. Mutual aid, not competition, was the key to evolution, he said. Genes were a metaphysical fiction; reductionism was a mistake. 'There is in an organism no special substance apart from the ordinary body ... We deny little pieces, corpuscles of heredity.' (After 1961 Russian scientists were allowed to study DNA, but Lysenko, in his confused way, made it clear that the double helix was a foolish notion: 'It deals with the doubling, but not the division of a single thing into its opposites, that is, with repetition, with increase, but not with development.')[18] Lysenkoism was an organic, 'holistic' science and a 'hymn to the natural union of men with their living environment'. It remained disdainful of demands for data to prove its claims, preferring bucolic folk wisdom.

Throughout the 1930s, the Lysenkoites fought an increasingly bitter battle for supremacy over the geneticists within Soviet biology. Gradually they gained the upper hand, and in 1948 Lysenko at last won full support from the state. Genetics was suppressed, geneticists were arrested and many died. The death of Stalin in 1953 made no difference, Khrushchev being an old friend and supporter of Lysenko. Yet it was increasingly obvious to Russian scientists – though not to many foreign biologists, who continued to apologise for Lysenko – that the man was a nutcase. Literally: he claimed to have created a hornbeam tree that bore hazelnuts, a wheat plant that grew rye seeds, and to have seen cuckoos hatching from warblers' eggs.

Lysenko fell with Khrushchev in 1964. Indeed, he was part of the reason Khrushchev fell. Lysenkoism was on the agenda of the meeting of the Central Committee that deposed Khrushchev, and the stagnation of agricultural yields since 1958 was the main charge against the party leader. Lysenko was disgraced, but the criticism was muted for many years. His science vanished without trace.[19]

NOTHING BUTTERY

This agricultural story may seem to have little to do with human nature. After all, as David Joravsky, a historian of Lysenkoism, has put it, 'any resemblance to genuinely scientific thought was purely accidental'. Yet it provides the background against which all Soviet biology operated. The extreme nurturism that began long before the revolution with Sechenov and reached its apogee under Lysenko set the tone for much of the century in Russia. And, consciously or not, it was echoed throughout the West. The insights of Pavlov and Watson into how learning occurred were somehow taken by many as proof that nothing but learning occurred in people. Marxism explicitly endorsed human exceptionalism, arguing that human history had switched from biology to culture at a specific moment ('Man, thanks to his mind, ceased long ago to be an animal,' said Lysenko). Marx was also credited with transcending the antinomy between 'is' and 'ought' – the famous naturalistic fallacy of David Hume and G.E. Moore. By the late 1940s the twin notions that human beings were the products of nurture and culture, in sharp contrast to animals, and that this was a moral as well as a scientific necessity, were widespread throughout the West as well as the socialist world.

'If genetic determinism is true,' wrote Stephen Jay Gould, 'we will learn to live with it as well. But I reiterate my statement that no evidence exists to support it, that the crude versions of past centuries have been conclusively disproved, and that its continued popularity is a function of social prejudice among those who benefit most from the status quo.'[20] This reasoning led to trouble. As biologists from Ernst Mayr to Steven Pinker have argued, it is not just mistaken to base policy and morality on an assumption of malleable human nature – it is dangerous. As soon as biologists began to discover that there was a degree of innate, genetic causation to behaviour, then another argument would have to be invented for morality. Said Pinker:

> Once [social scientists] staked themselves to the lazy argument that racism, sexism, war and political inequality were logically unsound or factually incorrect because there is no such thing as human nature (as opposed to morally despicable, regardless of the details of human nature), every discovery about human nature was, by their own reasoning, tantamount to saying that racism, sexism, war and political inequality were not so bad after all.[21]

I shall repeat myself just to be absolutely clear. There is nothing factually wrong with arguing that human beings are capable of learning, or being conditioned to associate stimuli, or reacting to reward and punishment or any other aspect of learning theory. These are true facts and vital bricks in the wall I am building. But it does not follow that therefore human beings have no instincts, any more than it would follow that human beings are incapable of learning if they have instincts. Both can be true. The error is to be a 100–nil person or a nil–100, to indulge in what the philosopher Mary Midgely calls 'nothing buttery'.

The high priest of nothing buttery was Burrhus Frederic Skinner, a follower of Watson, who took behaviourism to new heights of dogmatism. The organism, said Skinner, was a black box that need not be opened: it merely processed signals from the environment into an appropriate response, adding nothing from its innate knowledge. Skinner, even more than Watson, defined psychology by what was not true about human nature: that people did not have instincts. Even when, late in his life, he admitted that there was an innate component to human behaviour, he equated it with destiny – '[innate features] cannot be manipulated after the individual is conceived' – once again proving my point that the critics of innateness have a much more determinist model of genes in mind than the supporters. The nurturists were more fatalist about genes than the naturists.

I struggle to stay positive when reading Skinner. His experiments on operant conditioning were undoubtedly brilliant; his invention of the Skinner box, in which a pigeon could be rewarded or punished according to an experimental schedule, was a technological marvel; his intellectual honesty was undoubted. Unlike some behaviourists he did not pretend that environmentalism is not determinism. In my own life I frequently obey his tenets. I behave like a pigeon in a Skinner box when I go fly fishing: it was Skinnerians who discovered that an unpredictable random reward schedule is exceptionally effective in keeping the pigeon pecking at the symbol or the fisherman casting into the current. I behave like a Skinner box itself whenever I try to condition my children's table manners using reward and punishment.

Yet I cannot admire a man who regularly confined his own daughter Debby to a sort of Skinner box for the first two years of life. The 'air crib' was a soundproofed box with a window, supplied with filtered, humidified air, from which the little girl emerged only for scheduled playtimes and meals. Skinner also published a book attacking freedom and dignity as outmoded concepts. In 1948, the same year as George Orwell's *1984* appeared, he published a fictional account of utopia that sounds almost as bad as Orwell's hell. More of that later. My purpose here is to chart the decline and fall of Skinnerism, because it opened a new and fascinating chapter in the history of learning. It all began with a baby monkey in Wisconsin.

Harry Harlow was a jovial Midwestern psychologist addicted to puns and rhymes who chafed against the confines of his training in behaviourism. Born Harry Israel, he trained at Stanford under the dominating psychologist Lewis Terman (who insisted Harry change his name to Harlow because it sounded less Jewish and therefore improved his chances of getting a job). He never quite bought the idea that only reward and punishment determined the mind. Unable to build a rat laboratory he instead began rearing baby monkeys in a home-made laboratory when he moved to the University of Wisconsin at Madison in 1930. But soon he noticed that his baby monkeys, taken from their parents to be reared in perfect cleanliness and disease-free isolation, were growing up into fearful, anti-social, patently unhappy adults. They clung to cloths as if to rafts on the sea of life. One day in the late 1950s Harlow was on an aeroplane from Detroit to Madison when he looked down at the fluffy white clouds over Lake Michigan and was reminded of his baby monkeys clinging to their cloths. An idea for an experiment occurred to him. Why not offer a baby monkey the choice between a cloth model of its mother that did not reward it and a wire model of a mother that did reward it with milk? Which would it choose?

Harlow's students and colleagues were appalled by the idea. It was too fluffy a hypothesis for the hard science of behaviour. Eventually Robert Zimmerman was persuaded to do the experiment by the promise of being able to keep the baby monkeys for some more useful work later. Eight baby monkeys were placed in separate cages supplied with both wire-model mothers and cloth-model mothers – both were later equipped with lifelike wooden heads, mainly to please human observers. In four of the cages, the cloth mother contained a bottle of milk and a teat to drink from. In the other four, the milk came from the wire mothers. If these four baby monkeys had read Watson or Skinner they should quickly have come to associate the wire model with food and come to love wire. Their wire mothers rewarded them generously, whereas their cloth mothers ignored them. The baby monkeys spent nearly all their time on the cloth mothers; they would leave the security of the cloth only to drink from the wire mothers. In a famous photograph, a baby monkey clings with its rear legs to the cloth mother and leans across to get milk from a wire mother.[22]

Many similar experiments followed – rocking mothers were preferred to still ones, warm mothers to chilled ones – and Harlow announced the results in his presidential address to the American Psychological Association in 1958, provocatively entitling his talk 'The Nature of Love'. He had dealt a fatal blow to Skinnerism, which had talked itself into the absurd position that the entire basis of an infant's love for its mother was that the mother was the source of its nourishment. There was more to love than reward and punishment; there was something innate and self-rewarding about an infant's preference for a soft, warm mother. 'Man cannot live by milk alone,' quipped Harlow. 'Love is an emotion that does not need to be bottle- or spoon-fed.'[23]

There was a limit to the power of association, a limit supplied by innate preferences. These results seem almost absurdly obvious now, and to anybody who had read Tinbergen's work on the triggers of behaviour in gulls and sticklebacks they were obvious even then. But psychologists did not follow ethology, and such was the grip of behaviourism on psychology that Harlow's talk was genuinely surprising to many people. A crack had appeared in the edifice of behaviourism, a crack that would widen steadily.

Laboriously, throughout the 1960s, psychologists rediscovered the common-sense notion that people, and animals, are so equipped that they find some things easier to learn than others. Pigeons are rather good at pecking at symbols in Skinner boxes. Rats are good at running through mazes. By the late 1960s, Martin Seligman had developed the vital concept of 'prepared learning'. This was almost the exact opposite of imprinting. In imprinting, a gosling becomes fixated on the first moving thing it encounters, be it mother goose or professor: the learning is automatic and irreversible, but it can attach to a wide variety of targets. In prepared learning, the animal can learn to fear a snake very easily, for instance, but finds it hard to learn to fear a flower: the learning attaches only to a narrow range of targets, and without those targets it will not happen.

This fact was demonstrated by another group of Wisconsin monkeys a generation after Harlow. Susan Mineka was a student of Seligman, and after she moved to Wisconsin, in 1980, she designed an experiment to test the idea of prepared learning. She keeps the original videos of that experiment in a cardboard box in her office to this day. The clue that she followed up was the fact, known since 1964, that monkeys reared in the laboratory show no fear of snakes, whereas all wild-reared monkeys are scared witless by them. Yet it cannot be that every wild-reared monkey has had a bad Pavlovian experience with a snake, for the danger from snakes is usually lethal; you do not get much chance to learn by conditioning that snake bites are venomous. Mineka hypothesised that monkeys must acquire a fear of snakes vicariously, by observing the reactions of other monkeys to snakes. Lab-reared monkeys, not getting this experience, do not acquire the fear.

She first took six baby monkeys born in captivity to wild-born mothers and exposed them to snakes while they were alone. They were not especially afraid. When given the opportunity to reach over a snake to get some food, the hungry monkeys were quick to do so. Then she showed them snakes while their mother was present. The mother's terrified reaction – climbing to the top of the cage, smacking its lips, flapping its ears and grimacing – was immediately picked up by the offspring, which thereafter was permanently frightened even of a plastic model of a snake. (From now on, Mineka used toy snakes, which were easier to control.)

Next she showed that this lesson was just as easily learned from a strange monkey as from a parent, and then that it was easily passed on: a monkey could acquire a fear of snakes from a monkey that had acquired its own fear in this way. For her next trick, Mineka wanted to see if it was equally easy to get a monkey to teach a naïve monkey to fear something else, such as a flower. The problem was how to get the first monkey to react with fear to a flower. Mineka's colleague, Chuck Snowdon, suggested she use the newly invented technology of videotape. If monkeys could watch videotapes and learn from them, then the videos could be doctored to make it appear that the 'teaching' monkey was afraid of a flower, when it was in fact reacting to a snake.

It worked. Monkeys had no difficulty watching videotapes of monkeys and reacting to them as they did to real monkeys. So Mineka prepared tapes in which the bottom half of the screen was spliced in from another scene. This made it appear either that a monkey was calmly reaching over a model of a snake to get at some food, or that a monkey was reacting with terror to a flower. Mineka showed the doctored tapes to naïve lab-reared monkeys. In response to the 'true' tape (fear to snake, nonchalance to flower), monkeys quickly and robustly drew the conclusion that snakes are frightening. In response to the 'false' tapes (fear to flower, nonchalance to snake), monkeys merely drew the conclusion that some monkeys are crazy. They acquired no fear of flowers.[24]

This was, in my view, one of the great experimental moments in psychology, alongside Harlow's wire mother. It has been repeated in all sorts of different ways but the same conclusion always emerges clearly: monkeys very easily learn to fear snakes; they do not easily learn to fear most other objects. It shows that there is a degree of instinct in learning, just as imprinting shows that there is a degree of learning in instinct. Mineka's experiment has been much examined by blank-slate zealots desperate to find flaws in it, but so far it has resisted debunking.

Monkeys are not people, yet it is undoubtedly true that people are often afraid of snakes. Snake-fear is one of the commonest forms of phobia. Coincidentally, many people report that they gained their fear by a vicarious experience, such as seeing a parent react with fear to a snake.[25] People are also commonly afraid of spiders, the dark, heights, deep water, small spaces and thunder. All of these were a threat to Stone Age people, whereas the much greater threats of modern life – cars, skis, guns, electric sockets – simply do not induce such phobias. It defies common sense not to see the handiwork of evolution here: the human brain is pre-wired to learn fears that were of relevance in the Stone Age. And the only way that evolution can transmit such information from the past to the design of the mind in the present is via the genes. That is what genes are: parts of an information system that collects facts about the world in the past and incorporates them into good design for the future through natural selection.

Of course, I cannot prove the last few sentences. I can produce plenty of evidence that fear conditioning, in human beings as in other mammals, depends heavily on the amygdala, a small structure near the base of the brain.[26] I can even pass on a few hints about which servants of Vulcan are digging the trenches to and from the amygdala and how (it looks like the facilitation of glutamate synapses). I can tell you about twin studies showing that phobias are heritable, which implies genes at work. But I cannot be sure that all this is designed according to a plan laid out in a genetic instruction for wiring up the brain that way. I just cannot think of a better explanation. Fear learning looks like a clear-cut module, a blade on the mind's Swiss army knife. It is near-automatic, encapsulated, selective and operated by selective neural circuitry.

It still has to be learned. And you can also learn to fear cars, dentists' drills or sealskin coats. Clearly Pavlovian conditioning can create a fear of any kind. But it undoubtedly establishes a stronger, quicker and longer-lasting fear for snakes than for cars, and so can social learning. In one experiment, the human subjects were conditioned to fear snakes, spiders, electrical outlets or geometric shapes. The fear of snakes and spiders lasted much longer than the other fears. In another experiment, the subjects were conditioned (by loud bangs) to fear either snakes or guns. Again, the fear of snakes lasted longer than that of guns – even though snakes do not go bang.[27]

That a fear may be easily learned does not mean it cannot be prevented or reversed. Monkeys that have watched videos of other monkeys nonchalantly ignoring snakes become resistant to learning a fear of snakes even if later exposed to the video of an alarmed monkey. Children with pet snakes can apparently 'immunise' their friends against learning a fear of snakes. So this is not, Mineka stresses, a closed instinct. It is still an example of learning. But learning requires not just genes to set the system up for learning, but genes to operate it as well.

The most exciting thing about this story is the way it brings together each of the themes I have explored in this book so far. Superficially, fear of snakes looks exactly like an instinct. It is modular, automatic and adaptive. It is highly heritable – twin studies show that phobias, like personality, owe nothing to shared family environment, but a great deal to shared genes.[28] And yet – Mineka's experiments show it is entirely learnt. Was there ever a clearer case of nature via nurture? Learning is itself an instinct.

NERVES, NETS AND NODES

Hard-line behaviourists are rare birds these days. Few remain who have not been impressed by the cognitive revolution and by experiments like Mineka's into believing that the human mind learns what it is good at learning, and that learning requires more than a general-purpose brain; it requires special devices, each content-sensitive and each expert at extracting regularities from the environment. The discoveries of Pavlov, Thorndike, Watson and Skinner are valuable clues to how these devices go about their work, but they are not the opposite of innate: they depend on innate architecture.

There does remain a group of scientists who still object to injecting too much nativism into learning theory. They are called connectionists. As usual, what they actually say about how the brain works is barely distinguishable from what most nativists claim. But as usual, in nature-versus-nurture arguments the two sides of the argument like to paint each other into extreme corners, and feelings run high. The only difference I can see between the two is that the connectionists stress the openness of brain circuits to new skills and experiences while nativists stress their specificity. If you will forgive a bit of hack Latin, connectionists see the tabula as half rasa; nativists see it as half scripta.

To business, then. Connectionism is not really about real brains at all. It is about building computer networks that can learn. It gets its inspiration from two simple ideas: hebbian correlation and error back-propagation. The first refers to a Canadian named Donald Hebb, who made a throwaway remark in 1949 that has got him firmly into the history books:

> When an axon of cell A is near enough to excite cell B and repeatedly or persistently takes part in firing it, some growth process or metabolic change takes place in one or both cells such that A's efficiency as one of the cells firing B, is increased.[29]

What Hebb is saying is that learning consists of strengthening connections that are frequently in use. The servants of Vulcan dig out the channels that are used, making them flow better. Ironically, Hebb was no behaviourist – indeed, he was a fervent enemy of Skinner's idea that the black box must remain closed. He wanted to know what changed inside the brain, and he was right to guess that it was the strength of the synapse that changes. The phenomenon of memory, at the molecular level, seems to be precisely hebbian.

A few years after Hebb's insight, Frank Rosenblatt built a computer program called a perceptron, which consisted of two layers of 'nodes' or switches, the connections between which could be varied. Its job was to vary the strengths of the connections until its output had the 'correct' pattern. The perceptron achieved little, but when, 30 years later, a third, 'hidden', layer of nodes was added between the output and the input layers, the connectionist network began to take on the properties of a primitive learning machine, especially after being taught 'error back-propagation'. This means adjusting the strengths of the connections between the units in the hidden layer and the output layer where the output was in error, and then adjusting the strengths in the previous connections – propagating the error-correction back up the machine. It is broadly the same point about learning from prediction errors that modern Pavlovians make and that Wolfram Schultz found manifest in the human dopamine system.[30]

Connectionist networks, suitably well designed, are capable of learning regularities of the world in a manner that looks a bit like the way brains work. For instance, they can be used to categorise words into noun/verb, animate/inanimate, animal/human and so on. If damaged, or 'lesioned', they seem to make mistakes similar to those made by people who have had strokes. Little wonder that some connectionists are excited that they have taken the first steps to recreating the basic workings of the brain.

Connectionists deny that they believe in nothing but association. They do not, like Pavlov, claim that learning is a form of reflex, or like Skinner claim that a brain can be conditioned to learn anything with equal ease. Their hidden units play the innate role that Skinner was unwilling to grant the brain.[31] But they do claim that, with a minimum of prespecified content, a general network can learn a wide variety of rules about how the world works. In that sense they are in the empiricist tradition. They dislike excessive nativism, deplore the emphasis on massive modularity and are disgusted by cheap talk of genes for behaviour. Like David Hume, they believe that the knowledge the mind has derives largely from experience.

'That's what's so nice about empiricist cognitive science: you can drop out for a couple of centuries and not miss a thing,' quips the philosopher Jerry Fodor. Although Fodor has become a trenchant critic of taking nativism too far, he has no time for the connectionist alternative. It is 'simply hopeless', because it can neither explain what form logical circuits must take, nor explain the problem of abductive – 'global' – inference either.[32]

Steven Pinker's objection is more specific. He says that the achievements of connectionists are in direct proportion to the extent to which they pre-equip their networks with knowledge. Only by prespecifying the connections can you make a network learn anything useful. He compares connectionists to the man who claims to be able to make 'stone soup' – the more vegetables he adds, the better it tastes. In Pinker's view, the recent successes of connectionism are a backhanded compliment to nativism.[33]

In response, connectionists say they are not denying that genes may set the stage for learning, only that there may be general rules about how networks of synapses change to manifest that learning, and that similar networks may operate in different parts of the brain. They make much of recent discoveries of neural plasticity. In deaf people, or amputees, disused parts of the brain are reallocated to different functions, implying that they are multi-purpose. Speech, normally a left hemisphere function, is in the right hemisphere in some people. Violinists have a larger than usual somatosensory cortex for the left hand.

Far be it from me to referee such arguments. I would only make my usual judgement: something can be partly true without being the complete answer. I believe that there will be discovered networks in the brain that use their general properties as learning devices to learn regularities about the world, that they employ principles similar to connectionist networks and that similar networks may turn up in different mental systems so that learning to recognise a face uses a similar neuronal architecture as learning to fear a snake. Discovering those networks and describing their similarities will be fascinating work. But I also believe that there will be differences between networks that do different jobs, differences that encode pre-knowledge in the form of evolved design to a greater or lesser extent. Empiricists stress similarity; nativists stress difference.

Modern connectionists, like other empiricists before them – Hebb, Skinner, Watson, Thorndike, Pavlov, not to mention Mill, Hume and Locke – have undoubtedly added a useful brick to the wall. They are wrong only when they try to pull somebody else's bricks out, or to claim that the wall is held up only by empiricist bricks.

NEWTONIAN UTOPIA

Which brings me neatly back to Skinner. You will recall that he wrote a utopia. It describes as ghastly a place as Huxley's *Brave New World* or Galton's *Kantsaywhere*, and for the same reason: it is unbalanced. A world of pure empiricism untempered by genetics would be as terrible as a world of pure eugenics untempered by environment.

The book was called *Walden Two* and it is about a commune that is a suffocating cliché of fascism. Young men and women stroll through the corridors and gardens of the commune smiling and helping each other like something from a Nazi or Soviet propaganda film; the coercion of conformity is all around. No dystopian cloud mars the sky, and the hero, Frazier, is all the more creepy for the fact that his creator plainly admires him.

The novel is told through the eyes of a professor, Burris, who is taken by two former students to see his old colleague, Frazier, who has founded a community called Walden Two. Burris, accompanied by the students and their girlfriends plus a cynic called Castle, spends a week at Walden Two, admiring Frazier's apparently happy society based entirely on scientific control of human behaviour. Castle leaves, scoffing; Burris follows at first, but then returns, drawn back by the magnetism of Frazier's vision:

> Our friend Castle is worried about the conflict between long-range dictatorship and freedom. Doesn't he know he's merely raising the old question of predestination and free will? All that happens is contained in an original plan, yet at every stage the individual seems to be making choices and determining the outcome. The same is true of Walden Two. Our members are practically always doing what they want to do – what they 'choose' to do – but we see to it that they will want to do precisely the things which are best for themselves and the community. Their behavior is determined, yet they are free.[34]

I'm on Castle's side. But at least Skinner is honest. He sees human nature as entirely caused by outside influences, in a sort of Newtonian world of linear environmental determinism. If behaviourists were right, then the world would be like that: a person's nature would simply be the sum of external influences upon them. A technology of behaviour control would be possible. In a preface added to the second edition in 1976, Skinner shows few second thoughts, though like Lorenz he almost inevitably tries to tie Walden Two to the environmental movement.

Only by dismantling cities and economies, and replacing them with behaviourist communes, can we survive pollution, the exhaustion of resources and environmental catastrophe, says Skinner: 'Something like Walden Two would not be a bad start.' The truly scary thing is that Skinner's vision attracted followers who actually built a commune and tried to run it along Frazier's lines. It still exists: it is called Walden Dos and it is near Los Horcones in Mexico.[35]

C H A P T E R   E I G H T

Conundrums of culture

Some men by the unalterable frame of their constitutions, are stout, others timorous, some confident, others modest, tractable, or obstinate, curious or careless, quick or slow.

John Locke[1]

A child who comes into the world today inherits a set of genes and learns many lessons from experience. But she acquires something else, too: the words, the thoughts and the tools that were invented by other people far away or long ago. The reason the human species dominates the planet and gorillas are in danger of extinction lies not in our 5 per cent of special DNA, nor in our ability to learn associations, nor even in our ability to act culturally, but in the ability to accumulate culture and transmit information, across the seas and across the generations.

The word culture means at least two different things. It means high-brow art, discernment and taste: opera, in a word. It also means ritual, tradition and ethnicity: dancing round the camp fire with a bone through your nose. There is a deep convergence between the two: sitting in a black tie listening to *La Traviata* is merely a Western version of dancing round a camp fire with a bone through your nose. The first meaning of the word came out of the French Enlightenment. *La culture* meant civilisation – a cosmopolitan measure of progress. The second meaning came out of the German Romantic movement: *die Kultur* was the peculiar ethnic strain of Germanness that distinguished it from other cultures, the primeval essence of Teutonism. In England, meanwhile, arising out of the evangelical movement and its reaction to Darwinism, culture came to mean the opposite of human nature – the elixir that elevated man above the ape.[2]

Franz Boas, he of the magnificent moustaches in my imaginary photograph, brought the German usage to America and transmuted it into a discipline: cultural anthropology. His influence upon the nature–nurture debate of the ensuing century can hardly be exaggerated. By stressing the plasticity of human culture, he expanded human nature into an infinity of possibilities rather than a prison of constraints. It was he who most forcibly planted the idea that culture is what sets people free from their nature.

Boas's epiphany came on the shores of Cumberland Sound, a bay on the coast of Baffin Island in the Canadian Arctic. It was January 1884. Boas was 25 years old and he was mapping the coast to try to understand the migrations and the ecology of the Inuit people. He had recently switched interests from physics (his thesis was on the colour of water) to geography and anthropology. That winter, accompanied by only one European (his servant), he effectively became an Inuit: he lived with the Baffin Islanders in their tents and igloos, ate seal meat and travelled by dog sled. The experience was a humbling one. Boas began to appreciate not just the technical skills of his hosts, but the sophistication of their songs, the richness of their traditions and the complexity of their customs. He also saw their dignity and stoicism in the face of tragedy: that winter many Inuit died of diphtheria and influenza; their dogs, too, died by the score from a new disease. Boas knew the people blamed him for this epidemic. Not for the last time, an anthropologist would be left wondering if he had brought death to his subjects. As he lay in a cramped igloo listening to 'the shouting of the Eskimos, the howling of the dogs, the crying of the children', he confided to his diary: 'These are the "savages" whose lives are supposed to be worth nothing compared with a civilized European. I do not believe that we, if living under the same conditions, would be so willing to work or be so cheerful and happy!'[3]

In truth, he was well prepared for the lesson of cultural equality. He was the son of proudly free-thinking Jewish parents in the Rhineland town of Minden. His mother, a teacher, steeped him in 'the spirit of 1848', the year of Germany's failed revolution. At university he fought a duel to avenge an anti-Semitic slur, and bore the scars on his face for the rest of his life. 'What I want, what I will live and die for, is equal rights for all,' he wrote to his fiancée from Baffin Island. Boas was a fervent adherent of Theodor Waitz, who had argued for the unity of mankind: that all the races of the world descended from a recent common ancestor – a belief that split conservatives. It appealed to readers of Genesis disturbed by Darwin, but not to practitioners of slavery and racial segregation. Boas was also much influenced by the Berlin school of liberal anthropology of Rudolf von Virchow and Adolf Bastian, with its emphasis on cultural as opposed to racial determinism. So it was hardly a surprise when Boas concluded of his Inuit friends that 'the mind of the savage is sensible to the beauties of poetry and music, and that it is only to the superficial observer that he appears stupid and unfeeling'.[4]

Boas emigrated to the United States in 1887, and set about laying the foundations of modern anthropology as the study of culture, not race. He wanted to establish that the 'mind of primitive man' (the title of his most influential book) was every bit the equal of the mind of civilised man, and at the same time that the cultures of other people were deeply different from each other and from civilised culture. The origin of ethnic differences therefore lay with history, experience and circumstance, not with physiology and psychology. He first tried to prove that even the shapes of people's heads changed in the generation after they migrated to the United States:

The east European Hebrew, who has a very round head, becomes long-headed; the south Italian, who in Italy has an exceedingly long head, becomes more short-headed; so that in this country both approach a more uniform style.[5]

If the shape of the head – long a staple of racial taxonomy – was affected by the environment, then 'the fundamental traits of mind' could be, too. Unfortunately, a recent reanalysis of Boas's own data on skull shape suggest that it shows no such thing. Ethnic groups do retain distinct skull shapes even after assimilation into a new country. Boas's interpretation was influenced by wishful thinking.[6]

Though he stressed the influence of the environment, Boas was no extreme blank-slater. He made the crucial distinction between the individual and the race. It was precisely because he recognised profound innate differences in personality between individuals that he discounted innate differences between races: a perspective that was later proved genetically correct by Richard Lewontin. The genetic differences between two individuals chosen at random from one race are far greater than the average differences between races. Indeed, Boas sounds thoroughly modern in almost every way. His fervent anti-racism, his belief that culture determined rather than reflected ethnic idiosyncrasy and his passion for equality of opportunity for all would come to be hallmarks of political virtue in the second half of the century, when Boas was dead.

As usual, some of his followers went too far. They gradually abandoned Boas's belief in individual differences and his recognition of universal features of human nature. They made the usual mistake of equating the truth of one proposition with the falsehood of another. Because culture influenced behaviour, so innateness could not do so. Margaret Mead was initially the most egregious in this respect. Her studies of the sexual mores of Samoans purported to show how ethnocentric, and therefore 'cultural', was the Western practice of premarital celibacy and associated hang-ups about sex. In fact, it is now known she was duped by a handful of prank-playing young women during her all too brief visit to the island, and that 1920s Samoa was if anything slightly more censorious about sex than America.[7] The damage had been done, though, and anthropology, like psychology under Watson and Skinner, became devoted to the blank slate – to the notion that all of human behaviour was the product of the social environment alone.

In parallel with Boas's reformation of anthropology, the same theme was coming to dominate the new science of sociology. Boas's exact contemporary, and match in the moustache department, Emile Durkheim, made an even stronger statement of social causation: social phenomena could be explained by social facts alone, not by anything biological. *Omnia cultura ex cultura.* Durkheim was a year older than Boas, born in Lorraine, just across the French border from Boas's birthplace, also to Jewish parents. Unlike Boas, however, Durkheim was the son of a rabbi, descended from a long line of rabbis, and his youth was spent in the study of the Talmud. After flirting with Catholicism, he entered the elite École Normale Supérieure in Paris. Where Boas would wander the world, live in igloos, befriend native Americans and emigrate, Durkheim did little except study, write and argue. Aside from a brief period of study in Germany, he remained in the ivory tower of French universities all his life, first in Bordeaux and later in Paris. He is a biographical desert.

Yet Durkheim's influence upon the nascent school of sociology was immense. It was he who predicated the study of sociology on the notion of the blank slate. The causes of human behaviour – from sexual jealousy to mass hysteria – are outside the individual. Social phenomena are real, repeatable, definable and scientific (Durkheim envied the physicists their hard facts – physics envy is a well-known condition among softer scientists), but they are not reducible to biology. Human nature is the consequence, not the cause, of social forces.

> The general characteristics of human nature participate in the work of elaboration from which social life results. But they are not the cause of it, nor do they give it its special form; they only make it possible. Collective representations, emotions, and tendencies are caused not by certain states of the consciousnesses of individuals but by the conditions in which the social group, in its totality, is placed . . . Individual natures are merely the indeterminate material that the social factor molds and transforms.[8]

Boas and Durkheim, with Watson in psychology, represent the zenith of the blank-slate argument for the perfect malleability of human psychology by outside forces. As a negative rejection of all innateness, it is an argument that has been so demolished by Steven Pinker in his recent book *The Blank Slate* as to leave little to say.[9] But as a positive statement of the degree to which human beings are influenced by social factors, it is undeniable. The brick that Durkheim helped Boas put into the wall of human nature was a vital one – the brick called culture. Boas disposed of the notion that all human societies consisted of more or less well-trained apprentices to be English gentlemen, that there was a ladder of stages through which cultures must pass on the way to civilisation. In its place, he posited a universal human nature refracted by different traditions into separate cultures. The behaviour of a human being owes much to his nature; but it also owes much to the rituals and habits of his fellows. He seems to absorb something from the tribe.

Boas posed, and still poses, a paradox. If human abilities are the same everywhere, and Germans and Inuits have equal minds, then why are cultures diverse at all? Why is there not a single human culture common to Baffinland and the Rhineland? Alternatively, if culture, not nature, is responsible for creating different societies, then how can they be regarded as equal? The very fact of cultural change implies that some cultures can advance more than others, and if culture influences the mind, then some cultures must produce superior minds. Boas's intellectual descendants, such as Clifford Geertz, have answered the paradox by asserting that the universals must be trivial; there is no 'mind for all cultures', no common core to the human psyche at all save the obvious senses. Anthropology must concern itself with difference, not similarity.

This answer I find deeply unsatisfying, not least because of its obvious political dangers – without Boas's conclusion of mental equality, in by the back door comes prejudice. That would be to commit the naturalistic fallacy – deriving morals from facts, or 'ought' from 'is' – which the GOD forbid. It also commits the fallacy of determinism, ignoring the lessons of chaos theory: set rules need not produce a set result. With the sparse rules of chess, you can produce trillions of different games within just a few moves.

I do not believe Boas ever put it like this, but the logical conclusion of his position is that there is a great contrast between technological advance and mental stasis. Boas's own culture had steamships, telegraphs and literature; but it produced no discernible superiority in spirit and sensibility over the illiterate Inuit hunter-gatherers. This was a theme that ran through the work of Boas's contemporary, the novelist Joseph Conrad. Progress for Conrad was a delusion. Human nature never progressed, but was doomed to repeat the same atavisms in each generation. There is a universal human nature, retreading the triumphs and disasters of its ancestors. Technology and tradition merely refract this nature into the local culture: bow ties and violins in one place, nasal ornaments and tribal dancing in another. But the bow ties and the dances do not shape the mind – they express it.

When watching a Shakespeare play, I am often struck by the sophistication of his understanding of personality. There is nothing naïve or primitive about the way his characters scheme or woo; they are world-weary, jaded, postmodernist or self-aware. Think of the cynicism of Beatrice, Iago, Edmund or Jaques. I cannot help thinking, for a split second, that this seems odd. The weapons they fight with are primitive, their methods of travel cumbersome, their plumbing antediluvian. Yet they speak to us of love and despair and anger and betrayal in voices of modern complexity and subtlety. How can this be? Their author had such cultural disadvantages. He had not read Jane Austen or Dostoevsky; or watched Woody Allen; or seen a Picasso; or listened to Mozart; or heard of relativity; or flown in an aeroplane; or surfed the net.

Far from proving the plasticity of human nature, Boas's very argument for the equality of cultures depends upon accepting an unchanging, universal nature. Culture can determine itself, but it cannot determine human nature. Ironically, it was Margaret Mead who proved this most clearly. To find a society in which young girls were sexually uninhibited, she had to visit a land of the imagination. Like Rousseau before her, she sought something 'primitive' about human nature in the South Seas. But there is no primitive human nature. Her failure to discover the cultural determinism of human nature is the dog that failed to bark.

So turn the determinism around and ask why human nature seems to be universally capable of producing culture – of generating cumulative, technological, heritable traditions. Equipped with just snow, dogs and dead seals, human beings will gradually invent a lifestyle complete with songs and gods as well as sledges and igloos. What is it inside the human brain that enables it to achieve this feat, and when did this talent appear?

Notice, first, that the generation of culture is a social activity. A solitary human mind cannot secrete culture. It was the precocious Russian anthropologist Lev Semenovich Vygotsky who pointed out in the 1920s that to describe an isolated human mind is to miss the point. Human minds are never isolated. More than those of any other species, they swim in a sea called culture. They learn languages, they use technologies, they observe rituals, they share beliefs, they acquire skills. They have a collective as well as an individual experience; they even share collective intentionality. Vygotsky, who died at the age of 38 in 1934 after publishing his ideas only in Russian, remained largely unknown in the West until much later. But he has recently become a fashionable figure in educational psychology and some corners of anthropology. For my purposes, however, his most important insight is his insistence on a link between tool use and language.[10]

If I am to sustain my argument that genes are at the root of nurture as well as nature, then I must somehow explain how genes make culture possible. Once again, I intend to do so, not by proposing 'genes for' cultural practice, but by proposing the existence of genes that respond to the environment, of genes as mechanisms, not causes. It is a tall order and I may as well admit, right now, that I will fail. I believe that the human capacity for culture comes not from some genes that co-evolved with human culture, but from a fortuitous set of preadaptations that suddenly endowed the human mind with an almost limitless capacity to accumulate and transmit ideas. Those preadaptations are underpinned by genes.

THE ACCUMULATION OF KNOWLEDGE

The discovery that human beings are 95 per cent chimpanzee at the genetic level exacerbates my problem. In describing the genes involved in learning, instinct, imprinting and development, I had no difficulty calling on animal examples, for the difference between human and animal psychology in these respects is a difference of degree. But culture is different. The cultural gap between a human being and even the brightest ape or dolphin is a gulf. Turning an ancestral ape's brain into a human brain plainly took just a handful of minor adjustments to the recipe: all the same ingredients, just a little longer in the oven. Yet these minor changes had far-reaching consequences: people have nuclear weapons and money, gods and poetry, philosophy and fire. They got all of these things through culture, through their ability to accumulate ideas and inventions generation by generation, transmit them to others and thereby pool the cognitive resources of many individuals alive and dead.

An ordinary modern businessman, for instance, could not do without the help of Assyrian phonetic script, Chinese printing, Arabic algebra, Indian numerals, Italian double-entry book-keeping, Dutch merchant law, Californian integrated circuits, and a host of other inventions spread over continents and centuries. What is it that makes people, and not chimps, capable of this feat of accumulation?

After all, there seems little doubt that chimpanzees are capable of culture. They show strong local traditions in feeding behaviour, which are then passed on by social learning. Some populations crack nuts using stones; others use sticks. In West Africa, chimps eat ants by dipping a short stick into an ant's nest and putting each ant to the mouth one by one; in East Africa, they dip a long stick into an ant's nest, collect many ants on it and strip the ants off the stick into the hand and from there to the mouth. There are more than 50 known cultural traditions of this kind across Africa and each is learned by careful observation by youngsters (adult immigrants to a troop find it harder to learn local customs). These traditions are vital to their lives.

Frans de Waal goes so far as to say that 'chimps are completely dependent on culture for survival'. Like human beings, they cannot get through life without learned traditions.[11]

Nor are chimpanzees alone in this. The moment when animal culture was first discovered was in September 1953, on the tiny island of Kohima, off the coast of Japan. A young woman named Satsue Mito had for five years been feeding the monkeys on the islet with wheat and sweet potatoes to habituate them to human observers. That month she first saw a young monkey called Imo wash the sand off a sweet potato. Within three months two of Imo's playmates and her mother had adopted the practice and within five years most younger monkeys in the troop had joined them. Only the older males failed to take up the custom. Imo soon learn to separate wheat from sand by putting it in water and letting the sand sink.[12]

Culture abounds in large-brained species. Killer whales have traditional, and learned, feeding techniques that are peculiar to each population: beaching themselves to grab sea lions is a speciality of South Atlantic orcas, for instance, and a trick that requires much practice to perfect. So human beings are definitely not unique in being able to pass on traditional customs by social learning. But this only makes the question more baffling. If chimpanzees, monkeys and orcas have cultures, why have they not had cultural take-off? There is no ferment of continuous, cumulative innovation and change. There is, in a word, no 'progress'.

Rephrase the question, then. How did human beings get cultural progress? How did we happen on cumulative culture? This is a question that has elicited a torrent of theoretical speculation in recent years, but very little in the way of empirical data. The scientist who has tried hardest to pin down an answer is Michael Tomasello of Harvard. He has done a long series of experiments on adult chimpanzees and young human beings, from which he concludes that 'only human beings understand [other human beings] as intentional agents like the self and so only human beings can engage in cultural learning'. This difference emerges at nine months of age – what Tomasello calls the nine-month revolution. At this point human beings leave apes behind in the development of certain social skills. For instance, they will now point at an object for the sole purpose of sharing attention with another. They will look in the direction somebody points in, and follow the gaze of another. Apes never do this, and nor (until much later) do autistic children, who seem to have trouble with understanding that other people are intentional agents with minds of their own. According to Tomasello, no ape or monkey has ever shown the ability to attribute a false belief to another individual, something that comes naturally to most four-year-old human beings. From this, Tomasello infers that human beings can uniquely place themselves in others' mental shoes.[13]

This argument teeters on the brink of the human exceptionalism that so irritated Darwin. Like all such claims it is vulnerable to the first definitive discovery of an ape that acts on what it believes another ape is thinking. Many primatologists, not least Frans de Waal, feel they have already seen such behaviour in the wild and in captivity.[14] Tomasello will have none of it. Other apes can understand social relationships between third parties (something that is probably beyond most mammals) and they can learn by emulation. If shown that turning over a log reveals insects beneath, they will learn that insects can be found beneath logs. But they cannot, says Tomasello, understand the goals of other animals' behaviour. This limits their ability to learn, and in particular, it limits their ability to learn by imitation.[15]

I am not sure I buy Tomasello's full argument. I am influenced by Susan Mineka's monkeys, which are undoubtedly capable of social learning at least in the narrowly prepared case of snake fear. Learning is not some general mechanism; it is specially shaped for each kind of input and there may be inputs for which imitation learning is possible even in chimps. And even if Tomasello manages to explain away imitation in the cultural traditions of primates – the monkeys that learned to wash sand off potatoes, the chimps that learn from each other how to crack nuts – he will surely have trouble proving that dolphins cannot think their way into each other's thoughts. There is undoubtedly something uniquely human about the degree of our ability to empathise and imitate, just as there is something uniquely human about the degree of our ability to communicate symbolically – but it is a difference of degree, not kind.

Nevertheless, a difference of degree can still be amplified into a gulf by the ratchet of culture. Grant Tomasello his point that imitation becomes something more profound when the imitator has got inside the head of his model – when he has a theory of mind. Grant, too, that in some sense miming an idea to oneself creates representation, which in turn can become symbolism. Perhaps that is what enables young human beings to acquire much more culture than chimpanzees do. Imitation therefore becomes the first candidate part of what Robin Fox and Lionel Tiger called the culture acquisition device.[16] There are two other promising candidates: language and manual dexterity. And bizarrely, they all three seem to come together in one part of the brain.

In July 1991, Giacomo Rizzolatti made a remarkable discovery in his laboratory in Parma. He was recording from single neurons inside the brains of monkeys, trying to work out what causes a neuron to fire. Normally this is done in highly controlled conditions using largely immobile monkeys doing invented tasks. Dissatisfied with these artificial conditions, Rizzolatti wanted to record from monkeys leading almost normal lives. He began with feeding, trying to correlate each action with each neuronal response. He began to suspect that some neurons recorded the goal of the action, not the action itself, but his fellow scientists were dismissive: the evidence was too anecdotal.

So Rizzolatti put his monkeys back in a more controlled apparatus. From time to time each monkey was handed some food and Rizzolatti and his colleagues noticed that some 'motor' neurons seemed to respond to the sight of a person grasping a piece of food. For a long time they thought this was a coincidence and the monkey must be moving at the same time, but one day they were recording from a neuron which fired whenever the experimenter grasped a piece of food in a certain way; the monkey was completely still. The food was then handed to the monkey and as it grasped it in the same way, once again the neuron fired. 'That day I became convinced that the phenomenon was real,' says Rizzolatti. 'We were very excited.'[17] They had found a part of the brain that represented both an action and a vision of that action. Rizzolatti called it a 'mirror neuron' for its unusual ability to mirror both perception and motor control. He later found more mirror neurons, each active during the observation and imitation of a highly specific action: such as grasping between finger and thumb. He concluded that this part of the brain could match a perceived hand movement to an achieved hand movement. He believed he was looking at the 'evolutionary precursor of the human mechanism for imitation'.[18]

Rizzolatti and his colleagues have since repeated the experiment with human beings in brain scanners. Three bits of the brain lit up when the volunteers both observed and imitated finger movements: again, the phenomenon of 'mirror' activity. One of those areas was the superior temporal sulcus (STS), which lies in a sensory area concerned with perception. It is no surprise to find a sensory area lighting up when the volunteer observes an action, but it is surprising to find it active when the volunteer later executes the imitated action. A curiosity of human imitation is that if a person is asked to imitate a right-handed action, she will often imitate it with her left hand and vice versa. (Try telling somebody there is something on her cheek and touch your right cheek at the same time. Chances are, the person will touch her left cheek in response.) Consistent with this, in Rizzolatti's experiments, the STS was more active when the volunteer imitated a left-handed action with her right hand than when the volunteer imitated a left-handed action with the left hand. Rizzolatti concludes that the STS 'perceives' the subject's own action and matches it to its memory of the observed action.[19]

Recently, Rizzolatti's team has discovered a still stranger neuron, which fires not only when a certain motion is enacted and observed, but also when the same action is heard. For example, they found a neuron that responded to the sight and sound of a peanut being broken open, but not to the sound of tearing paper. The neuron responded to the sound of a breaking peanut alone, but not to the sight alone. Sound is important in telling the animal that it has successfully broken a nut, so this makes sense. But so exquisitely sensitive are these neurons that they can 'represent' certain actions from their sounds alone. This is getting remarkably close to finding the neuronal manifestation of a mental representation: the noun phrase 'breaking peanut'.[20]

Rizzolatti's experiments bring us close to describing, albeit in the crudest terms, a neuroscience of culture – a set of tools that between them make up at least part of the culture acquisition device. Will there be found a set of genes underlying the design of this 'organ'? In one sense, yes, for the content-specific design of brain circuits is undoubtedly inherited through the DNA. They may not be unique to this part of the brain, the uniqueness coming in the combination of genes used for the design rather than the genes themselves. These will create the capacity to absorb culture. But that is only one interpretation of the phrase 'culture genes'; a completely different set of genes from the designing genes will be found at work in everyday life. The axon-guidance genes that built the device will be long silenced. In their place will be genes that operate and modify synapses, that secrete and absorb neurotransmitters, and so on. Those will not be a unique set either. But they will in a true sense be the devices that transmit the culture from the outside world into and through the brain. They will be indispensable to the culture itself.

Recently, Anthony Monaco and his student Cecilia Lai discovered a genetic mutation apparently responsible for a speech and language disorder. It is the first candidate for a gene that may improve cultural learning through language. Severe Language Impairment has long been known to run in families, to have little to do with general intelligence and to affect not just the ability to speak, but the ability to generalise grammatical rules in written language and perhaps even to hear or interpret speech as well. When the heritability of this trait was first discovered, it was dubbed the 'grammar gene', much to the fury of those who saw such a description as guilty of determinism. But it now turns out that there is indeed a gene on chromosome 7, responsible for this disorder in one large pedigree and in another, smaller one. The gene is necessary for the development of normal grammatical and speaking ability in human beings, including fine motor control of the larynx. Known as forkhead box P2, or FOXP2 for short, it is a gene whose job is to switch on other genes – a transcription factor. When it is bust, the person never develops full language.[21]

Chimpanzees also have FOXP2; so do monkeys and mice. Merely possessing the gene does not therefore make speech possible. In fact, the gene is unusually similar in all mammals. Svante Paabo has discovered that in all the thousands of generations of mice, monkeys, orang-utans, gorillas and chimpanzees since they all shared a common ancestor, there have been only two changes in the FOXP2 gene that alter its protein product – one in the ancestors of mice and one in the ancestors of orang-utans. But perhaps having the peculiar human form of the gene is a prerequisite of speech. In human beings, since the split with chimpanzees (a mere yesterday) there have already been another two changes that alter the protein. And ingenious evidence from the paucity of silent mutations suggests that these changes happened very recently and were the subject of a 'selective sweep'. This is technical jargon for elbowing all other versions of the gene aside in short order. Some time after 200,000 years ago, a mutant form of FOXP2 appeared in the human race, with one or both of the key changes, and that mutant form was so successful in helping its owner to reproduce that his or her descendants now dominate the species to the utter exclusion of all previous versions of the gene.[22]

At least one of the two changes, which substitutes a serine molecule for an arginine at the 325th (out of 715) position in the construction of the protein, almost certainly alters the switching on and off of the gene. It might, for instance, allow the gene to be switched on in a certain part of the brain for the first time. This might, in turn, allow FOXP2 to do something new. Remember that animals seem to evolve by giving the same genes new jobs, rather than by inventing new genes. Admittedly, nobody knows exactly what FOXP2 does, or how it enables language to come into existence, so I am already speculating. It remains possible that rather than FOXP2 allowing people to speak, the invention of speech put pressure on the GOD to mutate FOXP2 for some unknown reason: that the mutation is consequence, not cause.

But since I am already beyond the perimeter of the known world, let me lay out my best guess for how FOXP2 enables people to speak. I suspect in chimpanzees the gene helps to connect the part of the brain responsible for fine motor control of the hand to various perceptual parts of the brain. In human beings, its extra (or longer?) period of activity allows it to connect to other parts of the brain including the region responsible for motor control of the mouth and larynx.

I think this because there may be a link between FOXP2 and Rizzolatti's mirror neurons. One of the parts of the brain active in the volunteers during Rizzolatti's grasping experiment, known as area 44, corresponds to the area where the mirror neurons were found in the monkey brain. This is part of what is sometimes called Broca's area, which thickens the plot considerably, because it is a vital part of the human brain's 'language organ'. In both monkeys and people, this part of the brain is responsible for moving the tongue, mouth and larynx (which is why a stroke in this area disables speech), but also for moving the hands and fingers. Broca's area does both speech and gesture.[23]

Herein lies a vital clue to the very origin of language itself. A truly extraordinary idea has begun to take shape in the minds of several different scientists in recent years. They are beginning to suspect that human language was originally transmitted by gesture, not speech.

The evidence for this guess comes from many different directions. First there is the fact that monkeys and people both use a completely different part of the brain to produce 'calls' from that which human beings use to produce language. The vocal repertoire of the average monkey or ape consists of several tens of different noises, some of which express emotions, some of which refer to specific predators and so on. All are directed by a region of the brain lying near the mid-line. This same region of the brain directs human exclamations: the scream of terror, the laugh of joy, the gasp of surprise, the involuntary curse. Somebody can be rendered speechless by a stroke in the temporal lobe and still exclaim fluently. Indeed, some aphasics continue to be able to swear with gusto, but find arm movements impossible.

The 'language organ', by contrast, sits on the (left) side of the brain, straddling the great rift valley between the temporal and frontal lobes –

the Sylvian fissure. This is a motor region, used in monkeys and apes mainly for gesture, grasp and touch, as well as facial and tongue movements. Most great apes are preferentially right-handed when they make manual gestures, and Broca's area is consequently larger on the left side of the brain in chimps, bonobos and gorillas.[24] This asymmetry of the brain – even more marked in human beings – must therefore have pre-dated the invention of language. Instead of the left brain growing larger to accommodate language, it would seem logical that language may have gone left because that was where the dominant gesturing hand was controlled. Nice theory, but it fails to explain the following ugly fact. People who learn sign language as adults do indeed use the left hemisphere; but native speakers of sign language use both hemispheres. Left-hemisphere specialisation for language is apparently more pronounced in speech than it is in sign language – the opposite of what the gesture theory predicts.[25]

A third hint in favour of the primacy of sign language comes from the human capacity for expressing language through the hands rather than the voice. To a greater or lesser extent people accompany much of their speech with gestures – even when speaking on a telephone, and even people who were blind from birth. The sign language used by deaf people was once thought to be a mere pantomime of gestures mimicking actions. But in 1960 William Stokoe realised that it was a true language: it uses arbitrary signs and it possesses an internal grammar every bit as sophisticated as spoken speech, with syntax, inflection and all the other accoutrements of language. It possesses other features very similar to spoken languages, such as being learned best during a critical period of youth and acquired in exactly the same constructive way as spoken languages. Indeed, just as spoken pidgins can be turned into fully grammatical creoles only when learned by a generation of children, so the same has proved true of sign languages. As final proof of the fact that speech is just one delivery mechanism for the language organ, deaf people can become manually 'aphasic' when they have strokes that affect the same regions of the brain as hearing people.

Then there is the fossil record. The very first thing that human ancestors did when they separated from the ancestors of chimps more than five million years ago was stand on their own two feet. Bipedal locomotion, accompanied by a massive reorganisation of the skeleton, occurred more than a million years before there was any sign of brain enlargement. In other words, our ancestors freed their hands to grasp and gesture long before they started to think or speak any differently from any other ape. One of the beauties of the gesture theory is that it immediately suggests why human beings got language and other apes did not. Bipedalism freed the hands not just to carry things, but to talk. The front limbs of most primates are too busy propping up the body to get into conversations.

Robin Dunbar suggests that language took over the role that grooming occupies in ape and monkey society – the maintenance and development of social bonds. Indeed, apes probably use their fine manual dexterity at least as much when seeking ticks in each other's fur as they do when picking fruit. In primates that live in large social groups, grooming becomes extremely time-consuming. Gelada baboons spend up to 20 per cent of their waking hours grooming each other. People started to live in such large groups, Dunbar argues, that it became necessary to invent a form of social grooming that could be done to several people at once: language. Dunbar notes that human beings do not use language just to communicate useful information; they use it principally for social gossip: 'Why on earth is so much time devoted by so many to the discussion of so little?'[26]

The grooming–gossip idea deserves an extra twist: if the first proto-humans to use language began to gossip with hand gestures, they would have necessarily neglected their real grooming duties. You can't groom and chew the fat at the same time if you talk with your hands. I am tempted to suggest that gestural language therefore brought with it a crisis of personal hygiene in our ancestors, which was solved only when they stopped being hairy and started wearing disposable clothes instead. But some waspish reviewer would accuse me of telling just-so stories, so I withdraw the idea.

According to the scanty fossil evidence, speech, unlike manual dexterity, appeared late in human evolution. The 1.6 million-year-old

Nariokotome skeleton discovered in 1984 in Kenya has in his neck vertebrae space for only a narrow spinal cord like an ape's, half the width of a modern human cord. Modern people need a broad cord to supply the many nerves to the chest for close control of breathing during speech.[27] Other still later skeletons of *Homo erectus* have high ape-like larynxes that might be incompatible with elaborate speech. The attributes of speech appear so late that some anthropologists have been tempted to infer that language was a recent invention, appearing as little as 70,000 years ago.[28] But language is not the same thing as speech: syntax, grammar, recursion and inflection may be ancient, but they may have been done with hands, not voice. Perhaps the FOXP2 mutation of less than 200,000 years ago represents not the moment that language itself was invented, but the moment that language could be expressed through the mouth as well as through the hands.

By contrast, the peculiar features of the human hand and arm appear early in the fossil record. Lucy, the 3.5 million-year-old Ethiopian, already had a long thumb and altered joints at the base of the fingers and in the wrist, enabling her to grasp objects between thumb, index and middle finger. She also had an altered shoulder allowing overarm throwing, and her erect pelvis allowed a rapid twist of the body axis. All three of these features are necessary for the human skill of grasping, aiming and throwing a small rock – something that is beyond the capability of a chimpanzee, whose throwing consist of randomly aimed, underarm efforts.[29] It is an extraordinary skill, requiring precision timing in the rotation of several joints and the exact moment of release. Planning such a movement requires more than a small committee of neurons in the brain; it needs coordination between different areas. Perhaps, says the neuroscientist William Calvin, it was this 'throwing planner' that found itself suited to the task of producing sequences of gestures ordered by a form of early grammar. This would explain why both sides of the Sylvian fissure, connected by a trunk line called the arcuate fasciculus, are involved.[30]

Whether it was throwing, tool-making or gesture itself that first enabled the perisylvian parts of the brain to become accidentally pre-adapted for symbolic communication, the hand undoubtedly played its part. As the neurologist Frank Wilson complains, we have too long neglected the human hand as a shaper of the human brain. William Stokoe, a pioneer of the study of sign language, suggested that hand gestures came to represent two distinct categories of word: things by their shape, and actions by their motion, so inventing the distinction between noun and verb that runs so deeply through all languages. To this day, nouns are found in the temporal lobe, verbs in the frontal lobe across the Sylvian fissure. It was their coming together that transformed a protolanguage of symbols and signs into a true grammatical language. And perhaps it was hands, not the voice, that first brought them together. Only later, perhaps to be able to communicate in the dark, did speech invade grammar. Stokoe died in 2000 shortly after completing a book on the hand theory.[31]

You can quibble with the historical details, and I am no diehard devotee of the hand–language hypothesis, but for me the beauty of this story lies in the way it brings imitation, hands and voice into the same picture. All are essential features of the human capacity for culture. To imitate, to manipulate and to speak are three things that human beings are peculiarly good at. They are not just central to culture: they are culture. Culture has been called the mediation of action through artefacts. If opera is culture, *La Traviata* is all about the skilful combination of imitation, voice and dexterity (in the making as well as the playing of musical instruments). What those three brought into being was a system of symbols, so that the mind could represent within itself, and within social discourse and technology, anything from quantum mechanics to the *Mona Lisa* or a motor car. But perhaps more important, they brought the thoughts of other minds together: they externalised memory. They enabled people to acquire far more from their social surroundings than they could ever hope to learn for themselves. The words, tools and ideas that occurred to somebody far away and long ago can be part of the inheritance of each individual person born today.

Whether the hand theory is right or not, the central role of symbolism in the expansion of the human brain is a proposition many can agree on. Culture itself can be 'inherited' and can select for genetic change to suit it. In the words of the three scientists most associated with this theory of gene–culture co-evolution:

> A culture-led process, acting over a long period of human evolutionary history, could easily have led to a fundamental reworking of human psychological dispositions.[32]

The linguist and psychologist Terence Deacon argues that at some point early human beings combined their ability to imitate with their ability to empathise and came up with an ability to represent ideas with arbitrary symbols. This enabled them to refer to ideas, people and events in their absence and so to develop increasingly complex culture, which in turn put pressure on them to develop larger and larger brains in order to 'inherit' items of that culture through social learning. Culture thereby co-evolves hand in hand with real genetic evolution.[33]

Susan Blackmore has even developed Richard Dawkins's idea of the meme to turn this process on its head. Dawkins describes evolution as competition between 'replicators' (usually genes) for 'vehicles' (usually bodies). Good replicators must have three properties: fidelity, fecundity and longevity. If they do, then competition between them, differential survival and hence natural selection for progressive improvement are not just likely – they are inevitable. Blackmore argues that many ideas and units of culture are sufficiently enduring, fecund and high-fidelity and that they therefore compete to colonise brain space. The words and concepts therefore provide the selection pressure to drive the expansion of the brain. The better brains were at copying ideas, the better they could cause their bodies to thrive.

> Grammatical language is not the direct result of any biological necessity, but of the way the memes changed the environment of genetic selection by increasing their own fidelity, fecundity and longevity.[34]

The anthropologist Lee Cronk gives a nice example of a meme. Nike, the shoe company, made a television advertisement featuring a group of East African tribesmen wearing Nike hiking boots. At the end of the commercial, one of the men turned to the camera and spoke some words. A subtitle translated: 'Just do it', the Nike slogan. Nike's luck was out, because the advert was seen by Lee Cronk, who speaks the Samburu dialect of Masai. What the man actually said was 'I don't want these. Give me big shoes.' Cronk's wife, a journalist, wrote the story up and it soon made it on to the front page of *USA Today* and into Johnny Carson's monologue on the *Tonight Show*. Nike sent Cronk a free pair of boots, which Cronk gave to a tribesman when next in Africa.

An everyday cross-cultural prank. It lasted a week in 1989 and was soon forgotten. But then a few years later, the Internet exploded on an unsuspecting world and Cronk's story soon found its way on to some website. From there it spread, minus the date, as if a new story, and Cronk now gets about one enquiry a month about it. The moral of the story is that memes need a medium to replicate in. Human society works quite well; the Internet works even better.[35]

As soon as human beings had symbolic communication, the cumulative ratchet of culture could begin to turn: more culture demanded bigger brains; bigger brains allowed more culture.

THE GREAT STANDSTILL

Yet nothing happened. Shortly after the time that the Nariokotome boy lived, 1.6 million years ago, there appeared on Earth a magnificent tool: the Acheulean handaxe. It was undoubtedly invented by members of the boy's species, the unprecedentedly huge-brained *Homo ergaster*, and it was a great leap forward from the simple, irregular, Oldowan tools that preceded it. Two-faced, symmetrical, shaped like a teardrop, sharpened all around, made of flint or quartz, it is a thing of beauty and mystery. Nobody knows for sure if it was used for throwing, cutting or scraping. It spread north to Europe with the diaspora of *Homo erectus*, the Coca-Cola of the Stone Age, and its technological hegemony lasted an incredible one million years: it was still in use just a half a million years ago. If this was a meme, it was spectacularly faithful, fecund and enduring. Yet astonishingly, during that time not one of the hundreds of thousands of people alive from Sussex to South Africa seems to have invented a new version. There is no cultural ratchet, no ferment of innovation, no experiment, no rival product, no Pepsi. There is only a million years of handaxe monopoly. The Acheulean Handaxe Corporation Inc. must have cleaned up. Big time.

Theories of cultural co-evolution do not predict this. They demand an acceleration of change once technology and language come together. The creatures that made these axes had big enough brains and versatile enough hands to make these handaxes, and to learn from each other how to do so, yet they did not use them to improve the product. Why did they wait more than a million years before suddenly beginning the inexorable, exponential progression of technology from spear-thrower to plough to steam engine to silicon chip?

This is not to denigrate the Acheulean handaxe. Experiments show that it is almost impossible to improve on it as a tool for butchering large game, except by inventing steel. It could only be perfected by the careful use of 'soft hammers' made of bone. But strangely, its makers seem to have had little pride in their tools, making fresh ones for each kill. In at least one case, at Boxgrove in Sussex, where more than 250 handaxes have been found, it appears that they were laboriously manufactured by at least six right-handed individuals at the site of a dead horse, then discarded nearby almost unused: some of the flakes knocked off in the process of making them showed more wear from butchery than the axes themselves. None of this explains why people capable of making such a thing did not also make spearheads, arrow points, daggers and needles.[36]

The writer Marek Kohn's explanation is that handaxes were not really practical tools at all, but the first jewellery: ornaments made by males showing off to females. Kohn argues that they show all the hallmarks of sexual selection; they are far more elaborate and (in particular) symmetrical than function demanded. They were artistry designed to impress the opposite sex, like the decorated bower built by a bower bird, or the elaborate tail grown by a peacock. That, says Kohn, explains the million years of stasis. Men were trying to make the ideal handaxe, not the best one. At least until very recently, in art and craft, Kohn argues, virtuosity, not creativity, has been the epitome of perfection. Women were obsessed with judging potential mates by their handaxe designs, not by their inventiveness. The image comes to mind of the maker of the best handaxe at Boxgrove sneaking off after a lunch of horse steaks for an assignation in the bushes with a fertile female, while his friends disconsolately pick up another lump of flint and start practising for the next occasion.[37]

Some anthropologists go further and argue that big-game hunting itself was sexually selected. For many hunter-gatherers, it was and is a remarkably inefficient way of getting food, yet men devote a lot of effort to it. They seem more interested in showing off by bringing back the occasional giraffe leg with which to entice a woman into sex than they are in filling the larder.[38]

I am a fan of the sexual selection theory, though I suspect it is only part of the story. But it does not solve the problem of the origin of culture; it is just a new version of brain–culture co-evolution. If anything, it makes the problem worse. The paleolithic troubadours whose ladies were so impressed by a well-crafted handaxe would surely have been even more impressed by a mammoth ivory needle or a wooden comb – something new. (Darling, I've got a surprise for you. Oh, honey, another handaxe: just what I always wanted.) Brains were growing rapidly bigger long before the Acheulean handaxe and they kept on getting bigger during the time of its long monopoly. If that expansion was driven by sexual selection, then why were the handaxes changing so little? The truth is that however you look at it, the mute monotony of the Acheulean handaxe stands in silent reproach over all theories of gene–culture evolution: brains got steadily bigger with no help from changing technology, because technology was static.

After half a million years, technological progress is steady, but very, very slow until the Upper Paleolithic Revolution, sometimes known as the Great Leap Forward. Around 50,000 years ago in Europe, painting, body adornment, trading over long distances, artefacts of clay and bone, elaborate new stone designs – all seem to appear at once. The suddenness is partly illusory, no doubt, because the toolkit had developed gradually in some corner of Africa before spreading elsewhere by migration or conquest. Indeed, Sally McBrearty and Alison Brooks have argued that the fossil record supports a very gradual, piecemeal revolution in Africa starting almost 300,000 years ago. Blades and pigments were already in use by then. They place the invention of long-distance trade at 130,000 years, for instance, based on the discovery at two sites in Tanzania of pieces of obsidian (volcanic glass) used to make spear points. This obsidian came from the Rift Valley in Kenya more than 200 miles away.

The sudden revolution of 50,000 years ago at the start of the Upper Paleolithic is clearly a Eurocentric myth, caused by the fact that far more archeologists work in Europe than in Africa. Yet there is still something striking to explain. The fact is that the inhabitants of Europe were culturally static until then, and so, before 300,000 years ago, were the inhabitants of Africa. Their technology showed no progress. After those dates, the technology changed with every passing year. Culture became cumulative in a way that it simply was not before. Culture was changing without waiting for genes to catch up.

I am faced with a stark and rather bizarre conclusion, one that I do not think has ever been properly confronted by theorists of culture and prehistory. The big brains which make people capable of rapid cultural progress – of reading, writing, playing the violin, learning about the siege of Troy, driving a car – came into being long before much culture had accumulated. Progressive, cumulative culture appeared so late in human evolution as to have had little chance to shape the way people think, let alone the size of their brains, which had already reached their maximum with little help from culture. The thinking, imagining and reasoning brain evolved at its own pace to solve the practical and sexual problems of life as a social species rather than to cope with the demands of culture transmitted from others.[39]

I am arguing that a lot of what we celebrate about our brains has nothing to do with culture. Our intelligence, imagination, empathy and foresight came into existence gradually and inexorably, but with no help from culture. They made culture possible, but culture did not make them. We human beings would probably be almost as good at playing, plotting and planning if we had never spoken a word, or fashioned a tool. If, as Nick Humphrey, Robin Dunbar, Andrew Whiten and others of the 'Machiavellian school' have argued, the human brain expanded to cope with social complexity in large groups – with co-operation, betrayal, deceit and empathy – then it could have done so without inventing language or developing culture.[40]

Yet culture does explain the ecological success of human beings. Without the ability to accumulate and hybridise ideas, people would never have invented farming, or cities, or medicine, or any of the things that enabled them to take over the world. The coming together of language and technology dramatically altered the fate of the species. Once they came together, cultural take-off was inevitable. We owe our abundance to our collective, not our individual brilliance.

Inexplicable as the origin of cumulative culture may be, once progress began, it fed upon itself. The more technologies people invented, the more food people could catch, the more minds those technologies could support and the more time people could spare for invention. Progress now became inevitable, a notion that is supported by the fact that cultural take-off happened in parallel in different parts of the world. Writing, cities, pottery, farming, currencies and many other things came together at the same time independently in Mesopotamia, China and Mexico. After four billion years with no literate culture, the world suddenly had three within a few thousand years or less. More if, as seems likely, Egypt, the Indus Valley, West Africa and Peru experienced cultural take-off independently. Robert Wright, whose brilliant book *Nonzero* explores this paradox in depth, concludes that human density played a part in human destiny. Once the continents were populated, albeit sparsely, and people could no longer emigrate to empty territory, density began to rise in the most fertile areas. With rising density came the possibility – nay, the inevitability – of increasing divisions of labour and therefore increasing technical invention. The population becomes an 'invisible brain' providing ever greater markets for individual ingenuity. And in those places where the available population suddenly shrank – such as Tasmania, when it was cut off from mainland Australia – cultural and technological progress did go suddenly into reverse.[41]

Density itself may not matter so much as what it allows: exchange. The prime cause of that success in the human species, as I argued in my book *The Origins of Virtue*, was the invention of the habit of exchanging one thing for another, for with it came the division of labour.[42] The economist Haim Ofek thinks it 'not unreasonable to view the Upper Paleolithic transition as one of the first in a series of fairly successful human attempts to escape (as populations) from poverty to riches through the institution of trade and the agency of the division of labor'.[43] He argues that what was invented at the start of the revolution was specialisation. Until that point, though there may have been sharing of food and tools, there was no allocation of different tasks to different individuals. The archeologist Ian Tattersall agrees: 'the sheer diversity of material production in [early modern human] society was the result of the specialization of individuals in different activities.'[44] Is it possible that once exchange and the division of labour were invented, progress was inevitable? There is certainly a virtuous circle at work in society today and has been since the dawn of history, whereby specialisation increases productivity, which increases prosperity, which allows technological invention, which further increases specialisation. As Robert Wright puts it, 'Human history involve[s] the playing of ever more numerous, ever larger and ever more elaborate non-zero-sum games.'[45]

So long as human beings lived, like other apes, in separate and competing groups, swapping only adolescent females, there was a limit to how rapidly culture could change, however well equipped human brains were to scheme, to woo, to speak or to think, and however high population density was. New ideas had to be invented at home; they could not generally be brought in. Successful inventions might help their owners to displace rival tribes and take over the world. But innovation came slowly. With the arrival of trade – exchange of artefacts, food and information initially between individuals and later between groups – all that changed. Now a good tool, or a good myth, could travel, could meet another tool or myth and could begin to compete for the right to be replicated by trade: in a word, culture could evolve.

Exchange plays the same role in cultural evolution that sex plays in biological evolution. Sex brings together genetic innovations made in different bodies; trade brings together cultural innovations made in different tribes. Just as sex enabled mammals to combine two good inventions – lactation and the placenta – so trade enabled early people to combine draught animals and wheels to better effect. Without exchange, the two would have remained apart. Economists have argued that trade is a recent invention, facilitated by literacy, but all the evidence suggests it is far more ancient. The Yir Yoront aborigines, living on the Cape York peninsula, were trading sting-ray barbs from the coast for stone axes from the hills through an elaborate network of trading contacts long before literacy.[46]

GENES THAT ALLOW CULTURE

All of this argument supports the conclusion that the progressive evolution of culture since the Upper Paleolithic revolution happened without altering the human mind. Culture seems to be the cart, not the horse; the consequence, not the cause, of some change in the human brain. Boas was right that you can invent any and every culture with the same human brain. The difference between me and one of my African ancestors of 100,000 years ago is not in our brains or genes, which are basically the same, but in the accumulated knowledge made possible by art, literature and technology. My brain is stuffed with such information, whereas his larger brain was just as stuffed but with much more local and ephemeral knowledge. Culture-acquiring genes do exist; but he had them too.

So what was it that changed about 200,000–300,000 years ago to enable human beings to achieve cultural lift-off in this way? It must have been a genetic change, in the banal sense that brains are built by genes and something must have changed in the way brains were built. I doubt it was merely a matter of size: a mutation in the ASPM gene allowing an extra 20 per cent of grey matter. More likely it was some wiring change that suddenly allowed symbolic or abstract thinking. It is tempting to believe it was FOXP2, which by rewiring the language organ somehow started the flywheel of exchange. But it seems just too fortunate for science to have stumbled on the key gene so early in its search, so I doubt that FOXP2 is the answer. I predict that the changes were in a small number of genes, simply because the lift-off is so sudden, and that before long science may know which ones.

Whatever the changes were, they enabled the human mind to take novelty in its stride much more than before. We are not selected to make minute predictive adjustments to a steering wheel while moving at 70 miles an hour, or to read handwritten symbols on paper, or to imagine negative numbers. Yet we can all do these things with ease. Why? Because some set of genes enables us to adapt. Genes are cogs in the machine, not gods in the sky. Switched on and off throughout life, by external as well as internal events, their job is to absorb information from the environment at least as often as to transmit it from the past. Genes do more than carry information; they respond to experience. It is time to reassess the very meaning of the word 'gene'.

SEX AND THE UTOPIA

If human nature did not change when culture changed – Boas's central insight, proven by archeology – then the converse is also true: cultural change does not alter human nature (at least not much). This fact has bedevilled utopians. One of the most persistent ideas in utopias is the abolition of individualism in a community that shares everything. Indeed, it is almost impossible to imagine a cult without the ingredient of communalism. The hope that the experience of a communal culture can change human behaviour flowers with special vigour every few centuries. From dreamers like Henri de Saint-Simon and Charles Fourier to practical entrepreneurs like John Humphrey Noyes and Bhagwan Shree Rajneesh, gurus have repeatedly preached the abolition of individual autonomy. The Essenes, the Cathars, the Lollards, the Hussites, the Quakers, the Shakers and the hippies have tried it, not to mention myriad sects too small to have memorable names. And there is one identical result: communalism does not work. Again and again, in accounts of these communities, what brings them down is not the disapproval of the surrounding society – though that is strong enough – but the internal tension caused by individualism.[47]

Usually, this tension first develops over sex. To condition human beings to enjoy free love and abolish their desire to be both selective and possessive about sexual partners seems to be impossible. You cannot even weaken this jealousy by rearing a new generation in a sharing culture: the jealous individualism actually gets worse in the children of the commune. Some sects survive by abolishing sex – the Essenes and Shakers were strictly celibate. This, however, leads to extinction. Others go to great lengths to try to reinvent sexual practice. John Noyes's Oneida community in upstate New York in the nineteenth century practised what he called 'complex marriage' in which old men made love to young women and old women to young men, but ejaculation was forbidden. In his Poona ashram, the Rajneesh initially seemed to have got free love going nicely. 'It is no exaggeration to say that we had a feast of f***ing, the likes of which had probably not been seen since the days of Roman bacchanalia,' boasted one participant.[48] But the Poona ashram, and the Oregon ranch that followed it, were soon torn apart by jealousy and feuds, not least over who got to sleep with whom. The experiment ended, 93 Rolls-Royces later, with attempted murder, mass food poisoning to gerrymander a local election and immigration fraud.

There are limits to the power of culture to change human behaviour.

CHAPTER NINE

The seven meanings of 'gene'

A scholar is just a library's way of making another library.
Daniel Dennett[1]

It is bad enough to be eclipsed on the brink of eternal fame by a competitor, but imagine how much worse it feels if that competitor has been dead for more than a decade and lived his entire life in total obscurity inside a monastery. No wonder Hugo De Vries stares rather unhappily out of my photograph. In 1900 he published a radical theory, for which he felt he deserved the sort of acclaim that had been showered upon John Dalton and was about to be showered on Max Planck. Where Dalton had suggested that matter is composed of atoms, and Planck would treat light as coming in lumps, De Vries too had come up with a quantum theory – that inheritance comes in particles: 'The specific characters of organisms are composed of separate units.'[2] He had deduced this by a series of brilliant experiments hybridising varieties of plants and he had even hit upon a truth that would take a century to be proven true. He speculated that the particles of heredity, which he called the 'pangens', did not obey the species barrier, so that a pangen for hairiness in one plant was also responsible for hairiness in another hairy species of flower.

De Vries, in other words, surely deserved to be known as the father of the gene. But soon after he rendered his triumphant account in print, in the French journal *Comtes Rendus de l'Académie des Sciences*, he was stung by a German bee called Karl Correns, a mild man driven into an uncharacteristic rage by De Vries's paper. Correns had been beaten to a scientific result by De Vries before and he was determined to have his revenge. Correns acidly pointed out that though De Vries's experiments were his own, his conclusion of particulate inheritance was borrowed, not just in outline but in detail, from the work of a long-dead Moravian monk named Gregor Mendel, even down to the terms De Vries used: recessive and dominant, for example.

Knowing he had been rumbled, De Vries conceded grudging priority to Mendel in a footnote to the German version of his paper, and settled unhappily for the role of rediscoverer of the laws of heredity. Worse, he had to share even this little credit with two other men: not only Correns, but also a young gate-crasher called Erich von Tschermark, who was good at only two things – persuading the world on flimsy evidence that he, too, had rediscovered Mendel's laws, and (much later) applying his talents in the service of Nazism. For De Vries, who had a high opinion of himself, this was bitter medicine; to the end of his days he looked on the deification of Mendel with disgust. 'This fashion is likely to pass,' he averred, refusing an invitation to the unveiling of a statue of the monk. The trouble was, not many people warmed to De Vries. Fastidious, aloof, tetchy and so misogynist that he was rumoured to spit in the culture plates of his female assistants, De Vries was doomed to see even his terminology eclipsed by that of others. By 1909 the pangen had become the 'gene', a word coined by Wilhelm Johannsen, a professor in Denmark.[3]

Was De Vries a plagiarist? Probably he did discover Mendel's laws through his own experiments before he rediscovered Mendel's work in the library: his sudden change of terminology in the late 1890s hints as much. In that sense, he made a great discovery. Probably, too, he thought he could get away with not citing Mendel's priority. After all, who read 40-year-old volumes of the Proceedings of the Brunn

Natural History Society for fun? In that sense, De Vries was a fraud. But it is no surprise when a scientist buries his ancestors, more or less unconsciously downplaying the insights of his predecessors lest they seem to diminish his own breakthrough. Even Darwin was adept, in his humble way, at skating over the contributions to his thinking of others, not least his own grandfather. Ironically, Mendel himself may have borrowed at least part of his main idea from another. He made no mention of the English horticulturist Thomas Knight's paper of 1799 showing how the easily achieved artificial pollination of different varieties of pea could hint at the mechanism of heredity, even down to the reappearance of characters in the second generation. Knight's paper, translated into German, was in the university library in Brunn.[4]

So, without taking anything away from Mendel, the irreplaceable genius of the gene, give De Vries his moment of glory as well. Let his concept of pangens, the interchangeable parts of heredity, stand for a moment alone and unique. Just as the different elements are made from different combinations of the same particles – neutrons, protons and electrons – so the world now knows, as it did not 20 years ago, that the different species are at least in part made from different combinations of very similar genes.

A GENE BY ANY OTHER NAME

During the twentieth century geneticists used at least five overlapping definitions of what a gene is. The first was Mendel's: a gene is a unit of heredity, an archive for the storage of evolutionary information. The discovery of the structure of DNA in 1953 immediately made Mendel's metaphor literal, by suggesting how genes could make genes. As James Watson and Francis Crick announced with arch understatement in *Nature*, 'It has not escaped our notice that the specific pairing we have postulated immediately suggests a possible copying mechanism for the genetic material.'[5] Merely by following the base-pairing rule that A must pair with T (and not C, G, or A), and that C must pair with G (and not C, T or A), each DNA molecule in two stages automatically produces an exact digital copy of its unique sequence. It needs a machine to do the copying, called DNA polymerase, but because the system is digital, it loses no precision, and because the system is fallible, it allows for evolutionary change. The Mendelian gene is an archive.

A second definition of the word gene, only recently revived, is De Vries's interchangeable part. The stunning surprise from the reading of genomes in the 1990s is that the human being has far more genes in common with the fly and the worm than anybody expected. The genes for laying down the body plan of the fruit fly turned out to have precise counterparts in the mouse and the man, all inherited from a common ancestor called the roundish flatworm that lived 600 million years ago. So similar are they that the human version of one of these genes can substitute for its fly counterpart in the development of a fruit fly. Even more surprising was the discovery that the genes flies use for learning and memory are also duplicated in people – and also presumably inherited from roundish flatworms. It is only a slight exaggeration to say that genes in animals and plants are a bit like atoms: standard-issue parts used in different combinations to produce different compounds. The De Vriesian gene is an interchangeable part.

A third definition of the gene starts in 1902 with De Vries's contemporary, the English doctor Archibald Garrod, who rather ingeniously identified the first single-gene disease, an obscure ailment called alkaptonuria. From him descends the all too common definition of genes by the diseases they cause when broken, the OGOD definition: one-gene-one-disease. This is misleading in two ways: it omits to mention that one mutated gene can be associated with many diseases, and one disease with many mutated genes; and it implies that the function of the gene is to prevent that disease. This is like saying the function of the heart is to prevent heart attacks. Still, given that most genetic research is driven by medical necessity, OGOD definitions are probably unavoidable. The Garrodian gene is a disease averter, a health giver.

A fourth definition of a gene is what it actually does. Right from the start, the DNA pioneers realised that genes had two jobs: copying themselves and expressing themselves through the construction of proteins. Garrod suggested that genes made enzymes: chemical catalysts. Linus Pauling broadened the point: genes made proteins of all kinds. Then, four months before the discovery of the double helix, James Watson suggested that DNA makes RNA, which makes protein, a concept later jauntily dubbed by Francis Crick 'the central dogma' of molecular biology. Information flows out of the gene and not back into it, just as information flows from the cook to the cake and not the other way. Though many details – alternative splicing, junk DNA, transcription factors, and most recently a plethora of new genes that make RNA but not protein, many of which seem to be intimately involved in regulating the expression of protein-coding genes – have complicated the standard picture of the metabolic gene, the central dogma still holds. With very few exceptions, protein does the work, DNA stores the information and RNA is the link between them, as Watson guessed. So the Watson–Crick gene is a recipe.

A fifth definition of the gene, which can be credited to the two Frenchmen François Jacob and Jacques Monod, is the gene as a switch and therefore as a unit of development. What Jacob and Monod did in the 1950s was to discover how a bacterium in a solution of lactose suddenly begins to produce the enzyme that enables digestion of lactose, and then stops making it when enough has been produced. The gene is switched off by a repressor protein, and the repressor is disabled by lactose. Jacob and Monod had guessed something like this must happen, floating the then startling idea that genes were turned on and off by the attachment of proteins to special sequences close to those genes; that, in other words, genes came with DNA switches. Now known as promoters and enhancers, these switches are the key to the development of a body from an embryo. Many genes require several activators to attach to their promoters; activators can work in different combinations; and some genes can be switched on by different sets of activators. The result is that the very same gene can be used in different species or in different parts of the body to produce completely different effects, depending on which other genes are also active. There is a gene called sonic hedgehog, for instance, which in one context turns neighbouring cells into neurons; in another context, it induces neighbouring cells to start growing into limbs. This is one reason that it is risky to speak of a 'gene for' something: many genes have multiple jobs.

Suddenly here is a very different way of viewing genes: as a set of developmental switches. All tissues carry the complete set of genes, but they are switched on in different combinations in different tissues. Now forget the sequence of the gene; what counts is where and how the gene is expressed. It is in this sense that many biologists now think of genes. To build a human body means throwing a series of switches in the right order, switches that cause growth and differentiation of the body. And just to make things interesting, the machines that throw the switches – the transcription factors – are themselves products of other genes. The Jacob–Monod gene is a switch.[6]

GENES WITH ATTITUDE

Yet, to tell the truth, there were legions of scientists who had been merrily using the word gene since it was coined in 1909 without really meaning any of these five concepts. For them, the gene was not the unit of heredity, evolution, disease, development or metabolism, so much as it was the victim of selection. It was Ronald Fisher who first clarified that evolution was little more than the differential survival of genes. And it was George Williams and William Hamilton, together with their bulldogs Richard Dawkins and Edward Wilson, who finally spelled out the full and startling implications of this idea. Bodies, said Dawkins, were temporary vehicles constructed for the replication of genes, exquisitely designed by genes to grow, to feed, to thrive, and die – but above all to strive for reproduction. Bodies were the genes' way of making new bodies. This 'gene's-eye view' of the organism was a sudden philosophical shift.

For instance, it immediately explains something that Aristotle, Descartes, Rousseau and Hume had not even realised needed explaining: why people are nice to their children (or in Rousseau's case, not). People are generally nicer to their own children than they are to other adults, other children and even to themselves. One or two twentieth-century anthropologists had feebly explained this in purely selfish terms – you are nice to your kids in the hope they will be nice to you in old age – but here, from Williams and Hamilton, was a genuine explanation that did not take the altruism out of parenting. You are nice to your children because you are descended from people who were nice to their children and were therefore better at enabling their children to survive to breed. This they could achieve because there are genes on their chromosomes that built their bodies in such a way that, given a certain environment, they would reliably produce behaviour in an adult that leads to reproduction and parental care. Targeted niceness could be in the genes.

Here is a definition of the word gene that is neither a unit of heredity, nor a unit of metabolism, nor a unit of development, but a unit of selection. It hardly matters for this purpose what this 'gene' is made of. It could be a pair of real genes, or a score. It could be a series of genes acting in sequence. It could be a network of genes, regulated by a plethora of RNAs. What counts is that it reliably produces a certain effect. How on earth does it do that? How can there be a gene that says 'Take care of your offspring!' in the language of DNA? And if there is such a gene, how can it thereby take care of itself? The whole concept – best known by Richard Dawkins's term 'The Selfish Gene' – seemed to many people almost magical. So used were they to thinking teleologically that they could not imagine a gene behaving selfishly unless it had the goal of selfishness in mind. Genes, asserted one critic, are just protein recipes; they 'cannot be selfish or unselfish, any more than atoms can be jealous, elephants abstract or biscuits teleological'.[7] But that was simply to miss Dawkins's point. For the sociobiologists, as they came to be called, the point was that natural selection could cause genes to act exactly as if guided by selfish goals: it was an analogy, but a remarkably useful one. People whose genes caused them, however indirectly, to be nice to their children left behind more descendants than people who did not.

It is now quite easy to build a link from the Watson–Crick gene to the Dawkinsian gene in real cases. Here is one, a gene on the northern tip of the Y chromosome called SRY. It is a tiny gene, just 612 letters long in a single exon (paragraph) of text – as simple as genes get. As a Mendelian unit of heredity, it replicates this 612-letter text. As a Watson–Crick unit of metabolism, it is translated into a 204-amino-acid protein called the testis-determining factor. As a Jacob–Monod unit of development it is switched on in parts of the brain and just one other tissue – the testis – for just a few hours, usually on the 11th day after conception (in mice). As a De Vriesian interchangeable pangen, it is found in much the same form in human beings as in mice and all mammals, where it performs a similar function – masculinising the body. As a Garrodian unit of disease, it is associated with various forms of sexual abnormality, most notably people with normal female bodies who none the less possess a Y chromosome, but lack a working version of this gene, or mice with normal male bodies who none the less possess no Y chromosome, but have a working version of this gene inserted into them by devious biologists. Broadly speaking, all an embryonic mammal needs to become a male is to have a single SRY gene, and to become a female it merely needs to lack a functioning version of the same gene.

For those readers who like to know how the engines of cars work, SRY probably performs this masculinising feat by one very simple action: it switches on another gene called SOX9. That is all it does. Genetically male human beings are occasionally born with one of their two SOX9 genes not working, and most of them develop into women with a skeletal disorder called campomelic dysplasia. SRY seems to be the captain of the ship casually ordering SOX9 to bring the vessel into port before retiring to its bunk. SOX9 does all the work, switching on and off all sorts of genes not only in the testis but in the brain as well – genes such as Lhx9, Wt1, Sf1, Dax1, Gata4, Dmrt1, Amh, Wnt4 and Dhh.[8] These genes in turn switch on and off the production of hormones, which alter the development of the body and in turn affect the expression of other genes. Many may prove to be sensitive to external experience, reacting to diet, social setting, learning and culture to refract the developing masculinity of the person. Yet it remains true that, given a typical middle-class upbringing, all the vast details of masculinity, as expressed in the modern environment – from testes to baldness to a tendency to sit on the couch drinking beer and flipping between channels on the television – stem from this single gene, SRY. It is surely not absurd to call it the gene 'for' maleness.

So you can easily see SRY as an archive, recipe, switch, interchangeable part or health-giver of maleness – depending on which of the twentieth century's five definitions of the word gene you prefer. You can just as easily see it as a unit of selection, a Dawkinsian selfish gene. Here's how. One of its downstream effects, inseparable from masculinity, is a greater likelihood that its body will take risks, act violently and die young. As soon as the testosterone of masculinity begins to bite in late adolescence, the premature mortality of males rises inexorably because of for main factors: homicides, suicides, accidents and heart disease. This is true even in Western societies – indeed, the gap between male and female mortality is widening. Of the major causes of death, only Alzheimer's kills more women than men. Nor is this some aberration of modern life. In some Amazon tribes more than half of the men are murdered. The average rate of violent death among men was higher in hunter-gatherer societies than it was in war-torn twentieth-century Germany.[9]

These risks are part of the package of being a man. Risk-taking is in the male essence – though it can be tempered by culture, varied by individuality and muted by technology. Old-fashioned Darwinian natural selection – the survival of the fittest individual – struggles to explain this fact. A gene whose consequence is higher mortality should head for rapid extinction. The reason it does not is obvious enough. Risk-averse wimps may live longer, but they do not have more children. The best way to reproduce, if you are a male, is to take a few risks, elbow a few other males out of the way and impress a few females. If you are lucky and have been born in middle-class California, you can do all this without much chance of actual death –

you may leave a few bruised egos and bent fenders behind, but you will probably survive. If you are less lucky and were born the son of a Yanomamo warrior, then your best bet for achieving genetic immortality is to kill and not be killed. In that society men who have killed other men have more sexual partners than average.[10] Whichever, there is no doubt that being a male is bad for survival and therefore fails the test of natural selection. The rational way out of this dilemma is to see the SRY gene, through the downstream effects of masculinising the body and brain, taking care of its own replication into future generations at the expense of the survival of its current body.

This is sexual selection, Darwin's other, much neglected theory, which urges not survival of the fittest, but reproduction of the fittest. Darwin regarded it as just as important as natural selection, perhaps more so in the case of human beings, but sexual selection spent most of the twentieth century in scientific exile. In its current form, as refined by people such as Amotz Zahavi and Geoffrey Miller, sexual selection theory suggests that the risk-taking of many male animals is an unconscious ploy by the genes of females to expose the genes of males to trial by fire so that she can be sure of selecting the best genes for her offspring. (In some species, it is the other way round.) Even if she passively watches males fighting over her, as seals and gorillas do, by mating with the winner she automatically selects fighting genes for future generations. Sexual selection of this kind can breed any type of male, from a vicious bully to a precious dandy to a gentle care-giver, and it can act upon the female, too, if exercised by the male. In socially monogamous species such as puffins or parrots, both sexes have bright colours to impress the other. In the human species, compared with other apes, there is clearly some degree of male selection for displaying youth, health, beauty and fidelity among females, while there is some female selection for displaying dominance, health, strength and fidelity among males.

A peahen that selects the male with the biggest, most ornamented train is unconsciously ensuring that the very act of growing a fancy tail is a test, a handicap that will reveal the quality of the male's genes. The more females express such a preference, the more males will inherit the capacity to grow the largest tails they can. To put it in corporate terms, peacock genes cannot be content with manufacturing a good body: they must market it. Like a toothpaste company, they have to put a lot into the advertising budget: the tail. Like an advertising budget, the tail seems a costly luxury, but it is vital. Such ornaments and rituals are, like advertising slogans, signals that try to be dishonest (does good toothpaste really improve your confidence?) but in the process help females honestly discriminate the genetic quality on offer in the mating market.

So Miller argues that it is no coincidence that many human talents – from storytelling to art, from jazz albums to sporting prowess to generosity to murder – tend to be displayed with greatest vigour by young male human beings at the age of mate selection. Miller points out that human beings devote ridiculous amounts of time to cultural practices that can only rarely enhance survival: art, dance, storytelling, humour, music, myth, ritual, religion, ideology. Yet all of these make sense as enhancers of reproductive success, of genetic rather than individual survival.[11]

Genes as units of instinct? The concept has travelled far from Mendel's hereditary particles. Confusion between many different conceptions of the gene has bedevilled the nature–nurture debate. You will no more find 'advertise male quality to females' written into the SRY gene than you will find 'advertise male wealth' written into the instruction manual of a Ferrari, but that does not mean it cannot be a valid interpretation of what each is for. Ferraris can be exquisite pieces of engineering at the same time as they can be sexual ornaments, and the same is true of genes.

ENTER POLITICS

This abstract concept of the Dawkinsian gene as a unit of instinct first hit the headlines in Edward O. Wilson's massive book on animal social behaviour, *Sociobiology*. Wilson, at Harvard, was an expert on the ecology of ants, and he was impressed, as all entomologists soon are, by the complexity of instinct. With no opportunity for learning, insects behave with sophistication and subtlety, but in a characteristic way for each species. The most striking aspect of ant behaviour is the way they delegate reproduction to a queen. Most ants, as workers, never breed. This fact had puzzled Darwin and it puzzled Wilson too, for it seemed to represent an exception to the rule that animals strive to reproduce. One day in 1965 Wilson boarded a train from Boston to Miami, having promised his wife he would not fly while their daughter was young. Trapped in the train for 18 hours, he turned to a new scientific paper by an obscure young British zoologist named William Hamilton. Hamilton had argued that the reason so many ants, wasps and bees were social was a quirk of their 'haplodiploid' genetics, which left workers more closely related to their sisters than to their daughters. So, in selfish gene terms, it paid them to raise the queen's offspring rather than their own. Hamilton's aim was broader than explaining ants – to draw attention to how such precise genetic calculus explains all cooperation between kin, the degree of instinctive co-operation being neatly related to the degree of relatedness. In other words, people are instinctively nice to their children because their genes make them that way, and their genes make them that way because genes that do so survive – through the children – at the expense of genes that do not.

Wilson at first found the paper naïve and foolish and tossed it aside after a cursory read, but he could not quite pin down its flaw. By the time his train was passing through New Jersey, he was rereading the paper more carefully. In Virginia he was frustrated and angry at Hamilton's presumption. Into northern Florida and Wilson was weakening. By the time he reached Miami, Wilson was a convert.[12]

Hamilton's theory – building on ideas from the self-effacing American, George Williams – dropped into the lives of many zoologists like a map into the lap of a lost explorer. Suddenly, they had a criterion by which to judge an explanation of an animal's behaviour: did it favour the propagation of its owner's genes? Richard Dawkins explored and expanded the implications of the idea in his beautiful book *The Selfish Gene*, but unlike Wilson he stuck to animals. Human beings, Dawkins said, were largely exceptions to the rule, because their conscious brains allowed them to ignore the dictates of their selfish genes.

Wilson had no such qualms. In the last chapter of *Sociobiology* he began to speculate about how human behaviours, too, might be the products of scheming genes. Was homosexuality a form of nepotism, genetically induced to allow childless 'uncles' to assist cooperative breeding? Did ethics need an evolutionary understanding? Could 'the social sciences shrink to specialised branches of biology'?[13] Wilson speculated 'in the free spirit of natural history', but at times he slipped into the evangelical language of the Baptist preachers of his Alabama youth. To the extent that he had a hidden agenda, he was motivated more by the desire to tweak the tail of religion than to fight for nature over nurture.[14] Indeed, he thought he was being mild and pluralist in his interpretation of how genes could collaborate with nurture to produce human social patterns. Aside from a few quasi-Marxist remarks about the inevitability of a planned society in the coming century, he had intended to say nothing overtly political. The storm that broke over his head in November 1975 took him genuinely by surprise.

It began with a letter to the *New York Review of Books* signed by a committee calling itself the Sociobiology Study Group. Among the 16 signatories were two of Wilson's Harvard colleagues and (he thought) friends: Stephen Jay Gould and Richard Lewontin. The letter accused Wilson of providing a new version of an old scheme,

> a genetic justification of the status quo and of existing privileges for certain groups according to class, race, or sex . . . Such theories provided an important basis for the enactment of sterilization laws and restrictive immigration laws by the United States between 1910 and 1930 and also for the eugenics policies which led to the establishment of gas chambers in Nazi Germany.[15]

As the controversy grew, spilling on to the cover of *Time* magazine the next year, it soon fell into the well-worn tracks of the nature–nurture debate, apparently pitting progressive but merciless environmentalists against conservative but hapless hereditarians. Wilson's lectures were picketed. Leaflets handed to students in Harvard Square accused him of postulating 'genes for all social life including war, business success, male supremacy and racism'.[16] Lewontin accused him of reflecting 'the ideologies of the bourgeois revolutions of the eighteenth century',[17] a standard term of abuse among Marxists. While he waited to respond to Gould at a symposium in Washington in 1979, Wilson was suddenly splashed with a glass of iced water by a group of chanting activists.

The argument was no less bitter across the Atlantic. Richard Dawkins, despite having largely ignored human beings in *The Selfish Gene* except to say that consciousness freed people from the tyranny of the genes, found himself accused of lending intellectual support to far-right politicians. Meanwhile, Wilson's attempts to explain himself at greater length, in two later books, persuaded some but largely failed to satisfy his critics, who were by now polarised into two extremes. He had encountered exactly the same wounded pride that had met Copernicus and Darwin: human beings do not enjoy seeing themselves removed from the centre of the universe. To see human behaviour dethroned from its supremacy and described in the same terms as ant behaviour was as insulting to the pride of the species as to see the Earth demoted to a planet. Perhaps also there would have been less vitriol if Wilson had talked about constellations of innate predispositions rather than 'genes'. The idea of a single sequence of DNA having the capacity to determine a human social attitude seemed intuitively wrong as well as humiliating.

Many biologists wedded to the selfish-gene view failed to come to Wilson's aid, causing a bitterness that lingers to this day. Some felt that Wilson's human speculations were naïve, premature and asking for trouble. Others were troubled by Wilson's imperialism: the boast that biology would soon take over the social sciences seemed at the very least insensitive. Others were merely in search of a quiet life: defending an alleged racist is to attract the label yourself. Indeed, a sharp division between genetically determined animals and culturally determined human being was a godsend for most biologists because it freed them:

to pursue their research in peace, without having to fear that they might accidentally stumble into or run afoul of highly charged social or political issues. It offers them safe conduct across the politicized minefield of modern academic life.[18]

The authors of this sentence, two other ex-Harvard scholars, John Tooby and Leda Cosmides, eschewing such safety, attempted a reform of sociobiology from within in 1992. They argued that the expressed behaviour of a human being need not be directly related to genes, but the underlying psychological mechanisms could be. So, to take a simple example, the search for 'genes for war' is bound to fail, but the contrary dogmatic insistence that war is a pure product of culture written on the blank slate of impressionable minds is equally foolish. There could well be psychological mechanisms in the mind, placed there by natural selection acting in the past upon sets of genes, that predispose most people to react to some circumstances in warlike ways. Tooby and Cosmides called this evolutionary psychology. It was an attempt to fuse the best of Chomsky's nativism – the idea that the mind cannot learn unless it has the rudiments of innate knowledge – with the best of sociobiology's selectionism: that the way to understand a part of the mind is to understand what natural selection designed it to do.

For Tooby and Cosmides it is the whole developmental program that evolves, the program for creating an eye, a foot, a kidney or a language organ in the brain. Each program requires the successful integration of hundreds, perhaps thousands of genes (many of them pangens used in other systems as well), and the presence of expected environmental cues. This is a subtle mixture of nature and nurture that studiously avoids pitting the two in opposition to each other:

Every time one gene is selected over another, one design for a developmental program is selected over another as well; by virtue of its structure, this developmental program interacts with some aspects of the environment rather than others, rendering certain environmental features causally relevant to development . . . Thus, both genes and the developmentally relevant environment are the product of natural selection.[19]

But crucially the environment is not some independent variable. The design of the developmental procedures specifies the environmental effects that will be used. Royal jelly turns a bee larva into a queen, but it does not turn a human baby into a queen. Genes, for Tooby and Cosmides, are designed to expect certain environments, and designed to make the most of them.

Despite this renewed emphasis on the environment, Tooby and Cosmides ran into the same political problem as Wilson and Dawkins. The social science establishment, liking their ambitions upon its subject matter no better than it had liked Wilson's, painted them as extreme reactionary nativists. I think this is a radical misinterpretation. For me, Tooby and Cosmides represent a retreat from naïve nativism towards an integration with nurture. The subject they helped to found – evolutionary psychology – is as comfortable with nurture explanations as it is with nature explanations. In the hands of Martin Daly and Margo Wilson, for example, it has been used to explain patterns of homicide and infanticide. Daly and Wilson recognise the role of sexual selection in making young adult males the prime perpetrators of murder, for example, but recognise just as strongly the role of the environment in producing the situations that actually elicit murder.[20] In the hands of Sarah Hrdy, evolutionary psychology has hypothesised that juvenile human beings are 'designed' by their past to expect to be reared in communal fashion, rather than in a nuclear family. It is impossible to parcel these studies into 'nature' or 'nurture'. They are about both. As Hrdy has put it:

Nature cannot be compartmentalised from nurture, yet something about human imaginations predisposes us to dichotomise the world that way . . . Complex behaviours like nurturing, especially when tied to even more complex emotions like 'love', are never either genetically predetermined or environmentally produced.[21]

The main complaint Tooby and Cosmides have against the social sciences is their desire to insulate themselves from other levels of explanation (to the cry of Reductionist!). Durkheim famously declared: 'Every time that a social phenomenon is directly explained by a psychological phenomenon, we may be sure that the explanation is false ... The determining cause of a social fact should be sought among the social facts preceding it and not among the states of individual consciousness.'[22] In other words, he rejected all reductionism. Yet other sciences have successfully integrated 'lower' levels of explanation without losing anything. Psychology uses biology, which uses chemistry, which uses physics. Tooby and Cosmides wanted to reinvent psychology in such a way that it used genes, not as implacable determinists of an inevitable human nature, but as subtle devices designed by ancestral selection to extract experience from the world.

The beauty of the Tooby–Cosmides gene, for me, is precisely this. It integrates all the other six definitions and adds a seventh. It is a Dawkinsian gene with attitude (in its dependence on passing the test of survival through the generations); a Mendelian archive (inscribed with the wisdom derived from millions of years of evolutionary adjustment); a Watson–Crick recipe (achieving its effects through the creation of proteins via RNAs); a Jacob–Monod developmental switch (expressing itself only in precisely specified tissues); a Garrodian health-giver (ensuring a healthy developmental outcome in the expected environment); and a De Vriesian pangen (reused in many different developmental programmes in the same species and in others). But it is also something else. It is a device for extracting information from the environment.

SRY, the masculinising gene on the Y chromosome, might seem at first glance to be a genetic determinist of the kind that gives social scientists the vapours. I have suggested that it sets in motion the sequence of events that (usually) leads to men sitting on couches drinking beer and watching football while women shop and gossip. But looked at another way it is the ultimate servant of nurture. Its job, aim and desire in life – with the help of hundreds of downstream genes – is to extract certain kinds of information from the upbringing and environment of its landlord organism. It extracts the food needed to grow a masculine body, the social cues needed to develop a masculine psyche, the gender cues needed to develop a masculine sexual preference, even the technology needed to express a masculine personality in the modern world (toy guns, say, or remote controls). It – or rather the developmental programme it starts – can be steered and adjusted by changes in that environment along the way. Take a baby boy from medieval Europe and time-transport him to modern California for his upbringing and it is a fair bet that his mind would be fascinated by guns and cars in place of swords and horses. SRY is no more than a glorified nurture-extractor.

Here again is the author's message of this book. Genes themselves are implacable little determinists, churning out utterly predictable messages. But because of the way their promoters switch on and off in response to external instruction, genes are very far from being fixed in their actions. Instead, they are devices for extracting information from the environment. Every minute, every second, the pattern of genes being expressed in your brain changes, often in direct or indirect response to events outside the body. Genes are the mechanisms of experience.

CHAPTER TEN

A budget of paradoxical morals

Why wrestle with Kant's God, Freedom, and immortality when it is only a matter of time before neuroscience, probably through brain imaging, reveals the actual physical mechanism that fabricates these mental constructs, these illusions?

Tom Wolfe[1]

When genes were discovered, late in the second millennium of the Christian era, they found a place already prepared for them at the table of philosophy. They were the fates of ancient myth, the entrails of oracular prediction, the coincidences of astrology. They were destiny and determination, the enemies of choice. They were constraints on human freedom. They were the gods.

No wonder so many people took against them. Genes got stuck with the label 'first cause'. Now that the genome is available for inspection, and genes can be seen at work, a much less terrifying picture is emerging. There are morals to be drawn from the nature–nurture debate, and in this chapter I intend to draw a few. They are mostly reassuring. The first and most general moral is that genes are enablers, not constrainers. They create new possibilities for the organism; they do not cut down its options. Oxytocin receptor genes allow pair-bonding; without them the prairie vole would not have the option of forming a pair bond. CREB genes allow memory; without those genes, it would be impossible to learn and recall. BDNF allows the calibration of binocular vision through experience; without it, you could not so easily judge depth and see the world as three-dimensional. FOXP2 mysteriously allows a human being to acquire the language of his people; without it, you cannot learn to speak. And so on. These new possibilities are open to experience, not scripted in advance. Genes no more constrain human nature than extra programs constrain a computer. A computer with Word, Powerpoint, Acrobat, Internet Explorer, Photoshop and the like cannot only do more than a computer without these programs – it can also get more from the outside world. It can open more files, find more websites and accept more emails.

Genes, unlike gods, are conditional. They are exquisitely good at simple if-then logic: if in a certain environment, then develop in a certain way. If the nearest moving object is a bearded professor, then that is what mothers look like. If reared in famine conditions, develop a different body type. Girls reared in fatherless households experience earlier puberty – an effect that is made possible by some still mysterious set of genes.[2] I suspect science has so far greatly underestimated the number of gene sets that act in this way – conditioning their output to external conditions.

So here is the first moral of the tale: *Don't be frightened of genes. They are not gods, they are cogs.*

MORAL NO. 2: PARENTS

Here comes another. In 1960 a graduate student at Harvard received a letter from George A. Miller, head of the department of psychology, dismissing her from the PhD programme because she was not up to the mark. Remember that name. Much later, stuck at home with chronic health problems, Judith Rich Harris took up writing psychology textbooks, books in which she faithfully relayed the dominant paradigm of psychology – that personality and much else was acquired from the environment. Then, 35 years after leaving Harvard, as an unemployed grandmother, having fortuitously escaped academic indoctrination, she sat down and wrote an article, which she submitted to the prestigious *Psychological Review*. It was published to sensational acclaim. She was deluged with curious enquiries as to who she was. In 1997 on the strength of the article alone she was given one of the top awards in psychology: the George A. Miller Award.[3]

The opening words of Harris's article were:

> Do parents have any important long-term effects on the development of their child's personality? This article examines the evidence and concludes that the answer is no.[4]

From about 1950 onwards psychologists had studied what they called the socialisation of children. Although they were initially disappointed to find few clear-cut correlations between a parenting style and a child's personality, they clung to the behaviourist assumption that parents were training their children's characters by reward and punishment, and the Freudian assumption that many people's psychological problems had been put there by their parents. This assumption became so automatic that to this day no biography is complete without a passing reference to the parental causes of the subject's quirks. ('It is probable that this wrenching separation from his mother was one of the prime sources of his mental instability,' says a recent author, referring to Isaac Newton.)[5]

To be fair, socialisation theory was more than an assumption. It did produce evidence, reams of it, all showing that children end up like their parents. Abusive parents produce abusive children; neurotic parents produce neurotic children; phlegmatic parents produce phlegmatic children; bookish parents produce bookish children, and so on.[6]

All of which proves precisely nothing, said Harris. Of course, children resemble their parents: they share many of the same genes.

Once the studies of twins raised apart started coming out, proving dramatically high heritability for personality, you could no longer ignore the possibility that parents had put their children's characters in place at the moment of conception, not during the long years of childhood. The similarity between parents and children could be nature, not nurture. Indeed, given that the twin studies could find almost no effect of shared environment on personality, the genetic hypothesis should actually be the null hypothesis: the burden of proof was on nurture. If a socialisation study did not control for genes, it proved nothing at all. Yet socialisation researchers went on year after year publishing these correlations without even paying lip-service to the alternative, genetic theory.

It was true that socialisation theorists used another argument as well: that different parenting styles coincide with different children's personalities. A calm home contains happy children; children who are hugged a lot are nice; children who are beaten a lot are hostile; and so on. But this could be confusing cause and effect. You could just as plausibly argue that happy children make a calm home; children who are nice get hugged a lot; children who are hostile get beaten a lot. Old joke: Johnny comes from a broken home; I'm not surprised – Johnny could break any home. Sociologists are fond of saying that a good relationship with parents 'has a protective effect' in keeping children off drugs. They are much less fond of saying that kids who do drugs do not get on with their parents.

So the correlation of good parenting with certain personalities is worthless as proof that parents shape personality because correlation cannot distinguish cause from effect. It is patent that socialisation, says Harris, is not something that parents do to children; it is something that children do to themselves. There is increasing evidence that what socialisation theorists have assumed were parent-to-child effects are often actually child-to-parent effects. Parents treat their children very differently according to the personalities of the children.

Nowhere is this more obvious than in the troubled matter of gender. Parents who are lucky enough to have children of different sexes will know that they treat them differently. They do not have to be told about the experiments in which adults rough-and-tumbled baby girls disguised in blue and cuddled baby boys disguised in pink. But most such parents will also hotly protest that the chief reason they treat their boys differently from their girls is because they are different. They fill the boy's cupboard with dinosaurs and swords, and the girl's with dolls and dresses, because they know that is the way to please each child. That is what the little tyrants keep asking for when in a shop. Parents may reinforce nature with nurture, but they do not create the difference. They do not force gender stereotypes down unwilling throats; they react to pre-existing prejudices. Those prejudices are not in one sense innate – there is no doll gene – but dolls are designed to appeal to predisposing prejudices, just as food is designed to appeal to human tastes. Besides, the parental reaction itself is just as likely to be innate: parents could be genetically predisposed to entrench rather than fight gender stereotypes.[7]

Once again, evidence for nurture is not evidence against nature, nor is the converse true. I just listened to a radio programme about whether boys were better at football than girls or whether their parents just pushed them that way. Protagonists for each view seemed implicitly to agree that their explanations were mutually exclusive. Nobody even suggested that both could be true at the same time.

Criminal parents produce criminal children – yes, but not if they adopt them. In a large study in Denmark, being adopted from an honest family into an honest family produced a child with a 13.5 per cent probability of getting into trouble with the law; that increased only marginally, to 14.7 per cent, if the adopting family included criminals. Being adopted from criminal parents to an honest family, however, caused the probability to jump to 20 per cent. Where both adopting and biological parents were criminals, the rate was even higher – 24.5 per cent. Genetic factors are predisposing the way people react to crimo-genic environments.[8]

Likewise, the children of divorced parents are more likely to divorce – yes, but only if they are biological children. Children whose adoptive parents divorce show no such tendency to follow suit. Twin studies reveal no role for the family environment at all in divorce. A fraternal twin has a 30 per cent probability of getting divorced if his twin gets divorced, about the same correlation as with a parent. An identical twin has a 45 per cent probability of divorce if his twin gets divorced. About half your divorce probability is in the genes; the rest is circumstance.

Rarely has an emperor seemed so naked after Harris was finished with socialisation theory. None of this will come as much surprise to people who have more than one child. Parenting is a revelation to most people. Having assumed you would now be the chief coach and sculptor of a human personality, you find yourself reduced to the role of little more than a helpless spectator-cum-chauffeur. Children compartmentalise their lives. Learning is not a backpack they carry from one environment to another; it is specific to the context. This is not to license parents to make their children unhappy – making another person suffer is wrong, whether it alters the person's personality or not. In the words of Sandra Scarr, the veteran champion of the idea that people pick the environments to suit their characters, 'Parents' most important job, therefore, is to provide support and opportunities, not to try to shape children's enduring characteristics.'[9] Sure, truly terrible parenting can still warp somebody's personality. But it seems likely that (I repeat) parenting is like vitamin C; as long as it is adequate, a little bit more or less has no discernible long-term effect.

Harris got brickbats as well as bouquets. In a long response, the authors of which included the doyenne of socialisation theory, Eleanor Maccoby, her critics surveyed studies supporting the notion that parents do affect personality, after all.[10] They conceded grudgingly that early socialisation theorists had exaggerated parental determinism, that twin studies needed to be considered and that a parent's behaviour is caused as much by the child's behaviour as vice versa. They emphasised that a criminal personality, even if partly genetic, is much more likely to be expressed in a criminal environment. And they drew attention to a series of studies demonstrating how drastically bad parenting could permanently affect a child. Romanian orphans adopted after the age of six months, for example, retain high levels of the stress hormone, cortisol, throughout their lives.

They also drew attention to the work of Stephen Suomi on rhesus monkeys. Suomi was a student of Harry Harlow who went on to build his own monkey laboratory at the National Institutes of Health in Maryland to continue Harlow's investigation of mother love. He first selectively bred monkeys to be highly strung. He then cross-fostered young monkeys to adoptive mothers for the first six months of their lives and studied their temperament and social life. A genetically nervous baby reared by a genetically nervous foster-mother turned into a socially incompetent adult, vulnerable to stress and itself a bad parent. But the same genetically jittery infant reared by a calm foster-mother – a 'supermom' – became quite normal, even rather good at rising to the top of the social hierarchy by making friends (sorry: 'recruiting social support') and evading stress. Despite its genetically nervous nature, such a monkey could become a calm and competent mother. Mothering style, in other words, is copied from the parent, rather than inherited.

Suomi's colleagues have since gone on to study the serotonin transporter gene in monkeys. One version of the gene produces a powerful and long-lasting reaction to maternal deprivation, whereas the other version of the gene is immune to maternal deprivation.[11] Since this gene also varies in human beings and the variation correlates with personality differences, this is a big finding. Translated into human terms it would imply that some children can be virtually orphaned and are none the worse for it; others need to be very well nurtured by their parents to turn out normal – the difference lies in the genes. Did we ever expect differently?

By citing Suomi's studies, Harris's critics show that they have already taken her lessons to heart: they are looking for how parents react to innate child personality and how they respond to genes. In their own words, they no longer see parents as 'moulding or determining' children. It is the nurturists who are calling for moderation, now. Gone is the triumphalism of Freud, Skinner and Watson. (Remember this? 'Give me a dozen healthy infants, well-formed, and my own specified world to bring them up in and I'll guarantee to take any one of them at random and train him to become any type of specialist

I might select – doctor, lawyer, artist, merchant-chief, and yes, even beggar-man and thief, regardless of his talents, penchants, tendencies, abilities, vocations and race of his ancestors.')

Moral: *Being a good parent still matters.*

MORAL NO. 3: PEERS

Harris's demolition of parental determinism is accompanied by the construction of an alternative theory. She believes the environment, as well as the genome, has a massive influence on the personality of a child, but mainly through the child's peer group. Children do not see themselves as apprentice adults. They are trying to be good at being children, which means finding a niche within groups of peers – conforming, but also differentiating themselves; competing, but also collaborating. They get their language and their accents largely from their peers, not their parents. Harris, like the anthropologist Sarah Hrdy, believes that ancestral human beings reared their children in groups, with women engaged in what zoologists call co-operative breeding. The natural habitat of the child was therefore a mixed crèche of children of all ages – almost certainly self-segregated by sex for much of the time. It is here, not in the nuclear family and the relation with parents, that we should look for the environmental causes of personality.

Most people think of peer pressure as pushing the young towards conformity. Seen from the balcony of middle age, teenagers seem obsessed with uniformity. Whether it be baggy, many-pocketed trousers, giant trainers, bare midriffs or backwards baseball caps, they prostrate themselves before the tyrant of fashion in the most craven way. Eccentrics are mocked; nonconformists ostracised. The code must be obeyed.

Conformity is indeed a feature of human society, at all ages. The more there is rivalry between groups, the more people will conform to the norms of their own group. But there is something else going on beneath the surface. Under the superficial conformity in tribal costumes lies an almost frantic search for individual differentiation. Examine any group of young people and you will find each playing a consistently different role. There is a tough, a wit, a brain, a leader, a schemer, a beauty. These roles are created, of course, by nature via nurture. Each child soon realises what he or she is good at and what he or she is bad at – compared with the others in the group. He then trains for that role and not for others, acting in character, developing still further the talent he has and neglecting the talent he lacks. The tough gets tougher, the wit gets funnier; and so on. By specialising in the role he has chosen, it becomes what he is good at. According to Harris this tendency to differentiate first emerges about the age of eight. Until that point, if a group of children is asked 'Who is the toughest boy here?', all will jump up crying 'Me!' After that age, they will start to say 'Him.'

This is true within families as well as in school classes and street gangs. The evolutionary psychologist Frank Sulloway sees each child within the family selecting a vacant niche. If the eldest child is responsible and cautious, the second child will often turn rebellious and carefree. Small differences in innate character are exaggerated by practice, not ironed out. Even among identical twins this happens. If one twin is more extroverted than the other, they will gradually exaggerate this difference. Indeed, psychologists find less correlation between fraternal twins in extroversion than there is between siblings of different ages: the very closeness in age causes them to exaggerate differences in personality. They are less alike than if they were two years apart. This is true of other measures of personality, too, and it seems to indicate a tendency for human beings to differentiate themselves from their closest companions by building on their innate propensities. If others are practical, then it pays to be cerebral.

I call this the Asterix theory of human personality. In the Goscinny and Uderzo cartoons about a defiant Gaulish village resisting the might of the Roman Empire, there is a very neatly drawn division of labour. The village contains a strong man (Obelix), a chief (Vitalstatistix), a druid (Getafix), a bard (Cacophonix), a blacksmith

(Fulliautomatix), a fishmonger (Unhygienix) and a man with bright ideas (Asterix). The harmony of the village owes something to the fact that each man respects the others' talents – with the exception of Cacophonix, the bard, whose songs are universally dreaded.

The first person to draw attention to this human tendency to specialise was probably Plato, but it was the economist Adam Smith who put the idea into circulation, and it was upon this observation that he built his theory of the division of labour – that the secret of human economic productivity is to divide labour among specialists and exchange the results. Smith reckoned human beings were unusual among animals in this respect. Other animals are generalists doing everything for themselves. Though rabbits live in social groups, there is no specialisation of function among them. No human being is truly a jack of all trades in the same way. Said Smith:

> In almost every other race of animals, each individual, when it is grown up to maturity, is entirely independent, and in its natural state has occasion for the assistance of no other living creature . . . Each animal is still obliged to support and defend itself, separately and independently, and derives no sort of advantage from that variety of talents with which nature has distinguished its fellows.[12]

But as Smith quickly went on to point out, specialisation is useless without exchange.

> Man has almost constant occasion for the help of his brethren, and it is in vain for him to expect it from their benevolence only. He will be more likely to prevail if he can interest their self-love in his favour, and show them that it is for their own advantage to do for him what he requires of them . . . It is not from the benevolence of the butcher, the brewer, or the baker, that we expect our dinner, but from their regard to their own interest. We address ourselves not to their humanity, but to their self-love, and never talk to them of our own necessities but of their own advantages. Nobody but a beggar chuses to depend chiefly upon the benevolence of his fellow-citizens.[13]

In this, Smith secured the backing of Emile Durkheim, who credited the division of labour as not just the source of social harmony but the foundation of the moral order as well:

> But if the division of labor produces solidarity, it is not only because it makes each individual an exchangist, as the economists say; it is because it creates among men an entire system of rights and duties which link them together in a durable way.[14]

I am intrigued by a coincidence: human adults are specialists, and human adolescents seem to have a natural tendency to differentiate themselves. Could it be that they are connected? In Smith's world, your adult speciality is a matter of chance and opportunity. You inherit the family bakery, perhaps, or you answer an advert for call-centre employees. You may be lucky and find a job that suits your temperament and talent, but most people just accept that they must learn to do the job they have. The role they played in an adolescent gang – as clown, raconteur, leader, tough – is long forgotten. Butchers, bakers and candlestick makers are made, not born. Or as Smith put it, 'The difference between the most dissimilar characters, between a philosopher and a street porter, for example, seems to arise not so much from nature, as from habit, custom and education.'

But human minds were designed for the Pleistocene savannah, not the urban jungle. And in that much more egalitarian world, where the same opportunities were open to all, talent may have decided your job. Imagine a band of hunter-gatherers. In the gang of youngsters playing round the camp fire are four adolescents. Og has just begun to notice that he has leadership qualities – he seems to be respected when he suggests a new game. Iz, on the other hand, has noticed that she can make the others laugh when she tells a story. Ob is hopeless with words, but when it comes to making a bark-strip net to catch rabbits he seems to have a natural talent. Ik, by contrast, is already a superb naturalist and the others are beginning to trust her to identify plants and animals. Over the next few years, each individual reinforces nature with nurture, specialising in one peculiar talent till it becomes a self-fulfilling prophecy. By the time they reach adulthood, Og no longer relies on natural talent for leadership; he has learnt it as a trade. Iz has practised the role of tribal bard so well it is second nature. Ob is even worse at making conversation, but he can now craft almost any tool. And Ik is a guru of lore and science.

The original genetic differences in talent may be very slight indeed. Practice has done the rest. But that practice may itself depend upon a sort of instinct. It is, I suggest, an instinct peculiar to human beings, deposited in the adolescent human brain by natural selection over tens of thousands of years, and it simply whispers in the ear of the juvenile: *Enjoy doing what you are good at; dislike doing what you are bad at.* Children seem to have this rule firmly in mind at all times. I am suggesting that the appetite for nurturing a talent might itself be an instinct. Having certain genes gives you certain appetites; finding yourself better at something than your peers sharpens your appetite for that thing; practice makes perfect and soon you have carved yourself a niche within the tribe as a specialist. Nurture reinforces nature.

So is musical or sporting ability nature or nurture? It is both, of course. Endless hours of practice is what it takes to play tennis or the violin well, but the ones who have an appetite for endless hours of practice are the ones with a slight aptitude and an appetite for practice. I recently had a conversation with the parents of a tennis prodigy. Was she always good at tennis? Not especially, but she was always keen to play, determined to join her elder siblings and badgering her parents for tennis lessons.

Moral: *Individuality is the product of aptitude reinforced by appetite.*

MORAL NO. 4: MERITOCRACY

As the last candidate left the room the chairman of the committee cleared his throat.

'Well, esteemed colleagues, we must choose one of those three people for the job of financial controller of the company: which is it to be?'

'Easy,' said the red-haired woman. 'The first one.'

'Why?'

'Because she is a qualified woman and this company needs more women.'

'Nonsense,' said the portly man. 'The best candidate was the second one. He has the best education. You can't beat Harvard Business School. Besides I knew his father at college. And he goes to church.'

'Pah,' scoffed the young woman with the thick glasses, 'When I asked him what seven times eight was, he said 54! And he kept missing the point of my questions. What use is a good education if you haven't a brain? I reckon the last candidate was by far the best. He was smooth, articulate, open and quick. He didn't go to college, true, but he's got a natural grasp of numbers. Besides, he's got a real personality and the chemistry's right.'

'Maybe,' said the chairman. 'But he's black.'

Question: who in this scene is guilty of genetic discrimination? The chairman, the red-haired woman, the portly man or the woman with glasses? Answer: all except the portly man. Only he is prepared to discriminate on the grounds of nurture. He is a true blank-slater, believing firmly that all human beings are born equal and stamped with their character by their upbringing. He is prepared to put his faith in the Church, Harvard and his college friend to create the right character whatever the raw material. The chairman's racism is based on the genetics of skin colour. The red-haired woman's adherence to affirmative action for women is discrimination against people with Y chromosomes. The young woman in the glasses prefers to ignore qualifications and look for intrinsic talent and personality. Her discrimination is more subtle, but it is certainly genetic, at least in part: personality is strongly inherited, and her dismissal of the Harvard candidate is based on the fact that his nurture genes have failed to take advantage of his education. She does not believe he is redeemable. I suggest she is just as much of a genetic determinist as the chairman and the red-haired woman – and of course I hope her candidate got the job.

Every job interview is about genetic discrimination. Even if the interviewer correctly ignores race, sex, disability and physical appearance, and discriminates on the grounds of ability alone, she is still discriminating, and unless she is prepared to go on qualifications and background alone – in which case, why hold an interview? – then she is looking for some intrinsic, rather than acquired, talent. The more she is prepared to make allowances for a deprived background, the more of a genetic determinist she is. Besides, the other point of the interview is to take into consideration personality, and remember the lesson of twin studies: personality is even more strongly heritable in this society than intelligence.

Do not misunderstand me. I am not saying it is wrong to interview people to try to ascertain their personality and their innate ability. Nor am I saying that it is right to discriminate on the grounds of race or genetic disability. Some forms of genetic discrimination are clearly more acceptable than others: personality is fine; race is not fine. I am saying that if you want to live in a meritocracy, then you had better not believe in nurture alone, or you will give all the top jobs to those who went to the top schools. The meritocracy means that universities and employers should select the best candidates despite – not because of – their backgrounds. And that means they must believe in inherited factors of mind.

Consider the question of beauty. You do not need a scientific study to tell you that some people are born more beautiful than others. Beauty runs in families; it depends on face shape, figure, nose size, and so on: all things that are mostly genetic. Beauty is nature. But it is also nurture. Diet, exercise, hygiene and accidents can all affect somebody's physical attractiveness, as can a haircut, make-up or cosmetic surgery. With plenty of money, luxury and help, even quite ugly people can make themselves attractive, as Hollywood proves regularly, and even beautiful people can ruin their looks with poverty, carelessness and stress. Some aspects of beauty show considerable cultural plasticity, notably thinness and fatness. In poor countries, and in the poorer past in the West, to be plump was to be beautiful and to be skinny was to be ugly; today, in the West, that equation has been at least partly reversed. Other aspects of beauty are less variable. If people from different cultures are asked to judge the beauty of women from photographs of their faces, a surprising degree of consensus emerges: Americans pick the same Chinese faces as Chinese people do; and Chinese pick the same American faces as Americans do.[15]

Yet how absurd it would be even to ask which aspects of beauty were nature and which nurture. Which bits of Britney Spears are genetically attractive and which are cosmetically attractive? It is a meaningless question, precisely because her nurture has enhanced rather than opposed her nature: her hairdresser has enhanced her hair, but it probably started out as quite nice hair. It is a fair bet, too, that her hair will be less attractive when she is 80 than when she is 20 due to – well, due to what? I was about to write some cliché like the ravages of the environment, and then I recalled that ageing is a largely genetic process, a process mediated by genes in the same way that learning is. The age-related decay of beauty that occurs in everybody after reaching adulthood is a process of nature via nurture.

There is heavy irony in the fact that the more egalitarian a society is, the more innate factors will matter. In a world where everybody gets the same food, the heritability of height and weight will be high; in a world where some live in luxury and others starve, the heritability of weight will be low. Likewise, in a world where everybody gets the same education, the best jobs will go to those with the most native talent. That's what the word meritocracy means.

Is the world more fair when every bright kid, even the one from the slums, gets a place at the best university, and so gets the best job? Is that fair on the stupid ones left behind? The message of the notorious book *The Bell Curve* was exactly this: that a meritocracy is not a fair place. Society stratified by wealth is unfair, because the rich can buy comforts and privileges. But society stratified by intelligence is also unfair, because the clever can buy comforts and privileges. Fortunately, the meritocracy is continually undermined by another, even more human force: lust. If clever men get to the top, it is a fair bet that they will use their privileges to seek out pretty women (and probably vice versa), just as the rich did before them. Pretty women are not necessarily stupid, but nor are they necessarily brilliant. Beauty will put the brake on stratification by brains.

Moral: *Egalitarians should emphasise nature; snobs should emphasise nurture.*

MORAL NO. 5: RACE

Seen from outside the species, human races look remarkably similar. To a chimpanzee or a Martian, the different ethnic groups of the human being would barely deserve classification as separate races at all. There are no sharp geographical boundaries where one race begins and the other ends, and the genetic variation between races is small compared with the genetic variation among individuals of the same race, reflecting the recent common ancestor of all human beings alive today – little more than 3,000 generations have passed since then.

But seen from the inside of one race, other human races look extremely different. White Victorians were ready to elevate (or relegate) Africans to a different species, and even in the twentieth century hereditarians frequently sought to prove that the differences between blacks and white were deeper than skin, and were manifest in the mind as well as the body. In 1972 Richard Lewontin disposed of most serious scientific racism by showing how genetic differences between individuals swamp those between races.[16] Though a few cranks still believe they will find justifications for racial prejudice in the genes, the truth is that science has done far more to explode than to foster the myth of racial stereotypes.

Yet racism has if anything moved up the political agenda even as racial prejudice and scientific justifications for it have faded. By the end of the century, sociologists were gingerly hinting at a new and disturbing idea – that however unjustified the science of race might be, racism itself might be in the genes. There might be an inevitable human tendency to be prejudiced against people of a different ethnic origin. Racism might be an instinct.

Ask Americans to describe another person they have only briefly met, and they will mention many features, perhaps including body weight, personality or hobbies. But three salient features will almost certainly get a mention: age, sex and race. 'My new neighbour is a young white woman.' It is almost as if it is one of the human mind's natural classifiers. The all-too-depressing conclusion is that if people are so naturally race-conscious, then maybe they are naturally racist.

John Tooby and Leda Cosmides refused to believe this. As the founders of evolutionary psychology, they are apt to think in terms of how instincts got started. Their logic was that back in the African Stone Age race was worse than useless as an identifier, because most people would never have met somebody of a different race. Noticing people's sex and age, on the other hand, would make good sense as these were reliable if approximate predictors of behaviour. So evolutionary pressures may well have built into the human mind an instinct – suitably transacted through nurture, of course – to notice sex and age, but not race. To them, it was a puzzle that race should therefore keep popping up as a natural classifier.

Perhaps, they reasoned, race is merely a proxy for something else. Back in the Stone Age – and before – one vital thing to know about a stranger is 'whose side is he on?' Human society, like all ape society, is riddled with factions – from tribes and bands down to temporary coalitions of friends. Perhaps race is merely a proxy for coalition membership. In other words, in modern America, people pay so much attention to race because they instinctively identify people of other races as being members of other tribes or coalitions.

Tooby and Cosmides asked their colleague Robert Kurzban to test this evolutionary theory by a simple experiment. It went like this. The subjects sat down at a computer and were shown a series of pictures each associated with a sentence putatively spoken by the person in the picture. At the end, they saw all eight pictures and all eight sentences, and they had to match each statement to the right picture. If they got them all right, Kurzban got no data: he was only interested in their mistakes. The mistakes told him something about how the subjects had classified people in their minds. For example, age, sex and race were, as expected, strong cues: the subjects would attribute a statement made by one old person to another old person, or one black person to another black person.

Now Kurzban started to introduce another possible classifier: coalition membership. This was revealed purely through the statements made by the depicted people, who were taking two sides of an argument. Quickly the subjects began to confuse two members of the same side more often than two members of different sides. Astonishingly, this largely replaced the tendency to make mistakes by race. But it had virtually no effect on the tendency to make mistakes by sex. Within four minutes, the evolutionary psychologists had done what social science had failed to do in decades: make people ignore race. The way to do it is to give them another, stronger cue to coalition membership. Sports fans are well aware of the phenomenon: white fans cheer a black player on 'their' team as he beats a white player on the opposing team.

This study has immense implications for social policy. It suggests that categorising individuals by race is not inevitable; that racism can be easily defeated if coalition cues cut across races; and that there is nothing intractable about racist attitudes. It also suggests that the more people of different races seem to act or be treated as members of a rival coalition, the more racist instincts they risk evoking. On the other hand, it suggests that sexism is a harder nut to crack because people will continue to stereotype men as men and women as women, even when they also see them as colleagues or friends.[17]

Moral: *The more we understand both our genes and our instincts, the less inevitable they seem.*

MORAL NO. 6: INDIVIDUALITY

I would hate to leave the reader feeling too comfortable. The discovery and dissection of genetic individuality is not going to make the life of politicians easier. Ignorance was once bliss; now they look back nostalgically to the time when they could treat everybody the same. In 2002, that innocence was lost for ever with the publication of an extraordinary study of 400 young men.

They were all born in the year 1972–73 in the city of Dunedin, on the South Island of New Zealand. Those born in that place and at that time were selected to be studied at regular intervals as they grew to adulthood. Of the 1,037 people in the cohort, Terrie Moffitt and Avshalom Caspi selected 442 boys who had four white grandparents. These children – all white and with little variation in class or wealth – included 8 per cent who were severely maltreated between the ages of 3 and 11 and 28 per cent who were probably maltreated in some way. As expected, many of the maltreated children have themselves turned out violent or criminal, getting into trouble at school or with the law and showing anti-social and violent dispositions. The nature-versus-nurture way to look at this would be to see whether this was because of the treatment they received from their abusive parents or the genes they received from those parents. But Moffitt and Caspi were interested in a nature-via-nurture approach instead. They tested the male children for differences in one particular gene called monoamine oxidase A, or MAOA, and then compared it with upbringing.

Upstream of the MAOA gene lies a promoter with a 30-letter phrase repeated 3, 3½, 4 or 5 times. Genes with the 3 and 5-repeat versions are much less active than those with 3½ or 4. So Moffitt and Caspi divided the young men into those with high-activity MAOA genes and those with low-activity MAOA genes. Remarkably, the ones with high-active MAOA genes were virtually immune to the effect of maltreatment. They did not get into trouble much even if maltreated as youngsters. Those with the low-active genes were much more anti-social if maltreated, and if anything slightly less anti-social than the average if not maltreated. The low-active maltreated men did four times their share of rapes, robberies and assaults.

In other words, it seems that it is not enough to have maltreatment, you must also have the low-active gene; or it is not enough to have the low-active gene, you must also be maltreated. The involvement of the MAOA gene comes as no great surprise. Knocking the gene out in a mouse causes aggressive behaviour, and restoring the gene reduces aggression. In a large Dutch family with a history of criminality over several generations, the MAOA gene was found to be broken altogether in the criminal family members and not in their law-abiding relations. But that mutation is very rare, and cannot explain much crime. The low-active, nurture-dependent mutations are much commoner (about 37 per cent of men).

The MAOA gene is on the X chromosome, of which males have only one copy. Women, having two copies, are correspondingly less vulnerable to the effect of the low-active gene, because most of them possess at least one version of the high-active gene as well. But 12 per cent of the girls in the New Zealand cohort did have two low-active genes, and these girls were significantly more likely to be diagnosed with conduct disorder when adolescent – if maltreated as youngsters.

Moffitt points out that reducing child abuse is a worthy goal whether it affects adult personality or not, so she sees no policy implications in the work. But it does not take much to imagine results like this opening the door to better intervention in the lives of troubled youngsters. It makes clear that a 'bad' genotype is not a sentence; it also requires a bad environment. Likewise, a 'bad' environment is not a sentence; it also requires a 'bad' genotype. For most people, the news is therefore liberating. But for a few it seems to slam the prison door of fate. Imagine you are a youngster rescued too late by the social services from an abusive family. Just one little diagnostic test, of the promoter length in this one gene, will allow a physician to predict, with some confidence, whether you are likely to be anti-social and probably criminal. How will you, your doctor, your social worker and your elected representative handle this knowledge? The chances are, that talk therapy would be useless, but that a drug to alter your mental neurochemistry would not: many drugs for mental conditions alter monoamine oxidase activity. But the drug could be risky, or might fail altogether. Politicians are going to have to decide who should have the power to authorise such a test and such a treatment, in the interests not only of the individual but of his future potential victims. Now that science knows the connection between gene and environment, ignorance is no longer morally neutral. Is it more moral to insist that all vulnerable people take such a test, to save them from future imprisonment, or that nobody be offered such a test? Welcome to the first of many Promethean dilemmas for the new century. Moffitt has already found another example of a genetic mutation in the serotonin system that responds to environmental factors. Watch this space.[18]

Moral: *Social policy must adapt to a world in which everybody is different.*

MORAL NO. 7: FREE WILL

When William James brought his considerable brain power to bear on the problem of free will in the 1880s, it was already a venerable conundrum. For all the efforts of Spinoza, Descartes, Hume, Kant, Mill and Darwin, he insisted that some juice still remained to be pressed from the free-will controversy. Yet even James was lamely reduced to the following disclaimer:

> I thus disclaim openly on the threshold all pretension to prove to you that the freedom of the will is true. The most I hope is to induce some of you to follow my example in assuming it is true.[19]

More than a century later, the same applies. For all the efforts of philosophers to impress upon the world that free will is neither an illusion nor an impossibility, the man and woman in the street are to all intents and purposes stuck where they were before. They can see the conundrum easily enough, but they cannot see the solution. To the extent that science posits a cause of someone's behaviour, it seems inevitably to take away his freedom of self-expression. Yet he feels he is free to choose his next act, in which case his behaviour is unpredictable. The behaviour is not random, though, so it must have a cause. And if behaviour has a cause, then it is not free. To all practical extent, the philosophers have failed to solve this problem in a way they can explain to the ordinary mortal. Spinoza said the only difference between a human being and a stone rolling down a hill is that the human being thinks he is in charge of his own destiny. Some help, that. Kant thought it inevitable that pure reason entangles itself in insoluble contradictions when trying to understand causality, and that escape lay through positing two different worlds, one run by the laws of nature, the other by intelligible agents. Locke said it was as nonsensical to ask 'whether a man's will be free as to ask whether his sleep be swift or his virtue square'. Hume said either our actions are determined, in which case there is nothing we can do about them; or our actions are random, in which case there is nothing we can do about them. Are we clear yet?[20]

I hope I have done enough in this book to convince you that appealing to nurture is no way out of the determinism dilemma. If personality is created by parents, peers or society at large, then it is still determined; it is not free. The philosopher Henrik Walter points out that an animal determined 99 per cent by genes and 1 per cent by its own agency has more free will than one determined 1 per cent by genes and 99 per cent by nurture. I hope, too, I have done enough to convince you that nature, in the shape of genes that influence behaviour, is no special or peculiar threat to free will. In some ways the news that your genes are important contributors to your personality should be reassuring: the very imperviousness of individual human nature to outside influences provides a bulwark against brainwashing. At least you are determined by your own intrinsic forces rather than somebody else's. As Isaiah Berlin put it, in the form of a catechism:

> I wish my life and my decisions to depend on myself, not on external forces of whatever kind. I wish to be the instrument of my own, not of other men's acts of will. I wish to be a subject, not an object.[21]

Incidentally, it is much bruited about that the discovery of genes influencing behaviour will lead to an epidemic of lawyers trying to excuse their clients on the ground that it was their genetic fate to commit crimes, not their choice. It was not his fault, Your Honour, it is in his genes. In practice, this defence has been tried in very few cases so far, and though it is bound to increase in frequency, I see no earth-shattering revolution in criminal justice if it does. For a start, the world is well used to deterministic excuses in the courts. Lawyers argue all the time for diminished responsibility on the grounds that the defendant was insane, was driven to it by her spouse, could not help himself because of the way he was treated as a child, and so on. Even Hamlet used the insanity defence in explaining to Laertes why he had killed his father, Polonius:

> What I have done,
> That might your nature, honour and exception
> Roughly awake, I here proclaim was madness.
> Was't Hamlet wrong'd Laertes? Never Hamlet:
> If Hamlet from himself be ta'en away,
> And when he's not himself does wrong Laertes,
> Then Hamlet does it not, Hamlet denies it.
> Who does it, then? His madness: if't be so,
> Hamlet is of the faction that is wrong'd;
> His madness is poor Hamlet's enemy.[22]

Genes are just another excuse to join the list. Besides, as Steven Pinker has pointed out, excusing criminals on the grounds of diminished responsibility is nothing to do with deciding whether they had free will to choose to behave as they did; it is merely about how to deter them from doing it again. But for me the chief reason the gene defence is still a rarity is that it is a pretty useless defence. In trying to disprove the charge against him, a criminal who admits to a natural inclination to crime is hardly likely to win over the jury. And when being sentenced, if he claims it is in his nature to murder, he is unlikely to persuade the judge to set him free to kill again. About the only reason for using the gene defence would be to avoid the death penalty after admitting guilt. The first case in which a genetic defence was used was indeed that of an Atlanta murderer, Stephen Mobley, appealing against the death penalty.

I am now going to attempt something much more ambitious: to convince you, as James could not, that the freedom of the will is true – despite nature and despite nurture. This is not to denigrate the great philosophers. Free will was, I believe, a genuinely insoluble problem until recent empirical discoveries, just as the nature of life was a genuinely insoluble problem until the discovery of the structure of DNA. The problem could not have been cracked by thought alone. It is probably still premature to tackle free will until we understand the brain better, but I believe we can now glimpse the beginning of a solution because of our understanding of what genes do in working brains.

Here goes. My starting point is the work of a visionary Californian neuroscientist with the appropriate name of Walter Freeman. He argues:

> The denial of free will, then, comes from viewing a brain as being embedded in a linear causal chain . . . Free will and universal determinism are irreconcilable boxes to which linear causality leads.[23]

The key word is linear, by which Freeman essentially means one-way. Gravity influences a falling cannon ball but not vice versa. Attributing all action to linear causality is a habit to which the human mind is peculiarly addicted. It is the source of many mistakes. I am not so concerned about the mistake of attributing cause where none exists, such as in the belief that thunder is Thor hammering, or in the search for blame for accidental events and the determinist obsession with horoscopes. My concern here is with another kind of mistake: the belief that intentional behaviour must have a linear cause. This is simply an illusion, a mental mirage, a misfiring instinct. It is quite a useful instinct, just as useful as the illusion that a two-dimensional image on a television screen is actually a three-dimensional scene. Natural selection has given the human mind a capacity for detecting intentionality in others, the better to predict their actions. We are fond of the metaphor of cause and effect as a means to understanding volition. But it is an illusion, all the same. The cause of behaviour lies in a circular, not a linear system.

This is not to deny volition. The capacity to act intentionally is a real phenomenon, and can be localised in the brain. It lies in the limbic system, as the following simple experiment demonstrates: an animal with any part of its forebrain cut off will lose a specific function. It will be blind, or deaf, or paralysed. But it will still be unmistakably intentional. An animal with its limbic system at the base of the brain excised is still perfectly capable of hearing, seeing and moving. If fed, it will swallow. But it initiates no action. It has lost its volition.

William James once wrote about lying in bed in the morning telling himself to get up. At first nothing happened; then, without noticing exactly how or when, he found himself getting up. He suspected that consciousness was somehow reporting the effects of the will, but was not the will itself. Since the limbic system is, roughly speaking, an unconscious area, this makes good sense. The decision to do something is made by your brain before you are aware of it. Benjamin Libet's controversial experiments with conscious epileptics seem to support the idea. Libet stimulated the brains of epileptics while they were under local anaesthetic. By stimulating the area of the left brain that receives sensory input from the right hand, he could make the patients consciously perceive a touch to the right hand, but only after half a second's delay. Then, by stimulating the left hand itself, he could get the same result plus an immediate, unconscious response in the appropriate part of the right brain, which had received its stimulus from the hand by a more direct, faster nerve. Apparently the brain can receive and start acting upon the sensation in real time before the inevitable delay required to process the sensation into awareness. This suggests that volition is unconscious.

For Freeman, the alternative to linear causality is circular causality, in which an effect influences its own cause. This removes the agency from the action, because a circle has no beginning. Imagine a flock of birds twisting and turning as it flies along the seashore. Each bird is an individual taking its own decisions. There is no leader. Yet they seem to turn in unison as if linked with one another. What is the cause of each twist and turn? Put yourself in the position of a single bird. You turn left, which causes your neighbour to bank to the left almost instantaneously. But you turned because your other neighbour turned, and he turned because he thought you were turning before you were. This time the little manoeuvre peters out because all three of you correct your path on seeing what the rest of the flock is doing, but next time perhaps the entire flock may catch the habit and swerve left. The point is, you will search in vain for a linear sequence of cause and effect, because the first cause (you appearing to turn) is then dramatically influenced by the effect (the neighbour turning). Causes can still only go forward in time, but they can then influence themselves. Human beings are so obsessed by linear causes that they find it almost impossible to escape the habit. We invent absurd myths, like the flap of a butterfly's wing starting a hurricane, in a vain attempt to preserve linear causality in such systems.

Freeman is not the only one to champion non-linear causality as the source of free will. The German philosopher Henrik Walter believes that the full ideal of free will is genuinely an illusion, but that people do possess a lesser form of it, which he calls natural autonomy and which derives from the feedback loops within the brain, where the results of one process become its next starting conditions. Neurons in the brain are hearing back from their recipients even before they have finished sending messages. The response alters the message they send, which in turn alters the response and so on. This idea is fundamental to many theories of consciousness.[24] Now try to imagine this in a parallel system with many thousands of neurons communicating at once. You will not get chaos, just as you do not get chaos in the bird flock, but you will get sudden transitions from one dominant pattern to another. You are lying awake in bed and the brain is freewheeling from one idea to another in that rather pleasant way. Each idea comes unbidden because of its associations with the last, as a new pattern of neuronal activity comes to dominate consciousness; then suddenly a sensory pattern intervenes – the alarm clock. Another pattern takes over (*I must get up*), then another (*Maybe a few minutes more*). Then before you know it a decision is taken somewhere in the brain and you become aware you are getting up. It is plainly a volitional act, yet it is in some sense determined by the alarm clock. To try to find the first cause of the actual moment of rising would be impossible, because it is buried in a circular process in which thoughts and experiences fed off each other.

Even the genes themselves are steeped in circular causality. By far the most important discovery of recent years in brain science is that genes are at the mercy of actions as well as vice versa. The CREB genes that run learning and memory are not just the cause of behaviour; they are also the consequence. They are cogs responding to experience as mediated through the senses. Their promoters are designed to be switched on and off by events. And what are their products? Transcription factors – devices for switching on the promoters of other genes. Those genes alter the synaptic connections between neurons; this in turn alters the neural circuitry, which in turn alters the expression of the CREB genes by absorbing outside experience, and so on round the circle. This is memory, but other systems in the brain are going to prove to be similarly circular. Senses, memory and action all influence each other through genetic mechanisms. These genes are not just units of heredity – that description misses the point of them altogether. They are themselves exquisite mechanisms for translating experience into action.[25]

I cannot pretend I have given a fine-grained description of free will, because I think none can yet exist. It is the sum and product of circular influences with varying networks of neurons, immanent in a circular relationship between genes. In Freeman's words, 'each of us is a source of meaning, a wellspring for the flow of fresh constructions within our brains and bodies'.

There is no 'me' inside my brain; there is only an ever-changing set of brain states, a distillation of history, emotion, instinct, experience and the influence of other people – not to mention chance.

Moral: *Free will is entirely compatible with a brain exquisitely prespecified by, and run by, genes.*

EPILOGUE

Homo stramineus – the straw man

Dead men tell no tales, and if there were any tribes of other type than this they have left no survivors. Our ancestors have bred pugnacity into our bone and marrow, and thousands of years of peace won't breed it out of us.

William James[1]

Twelve hairy men posed for my imaginary photograph in 1903. Had they met, I doubt whether they would have much liked each other. Abrasive Watson, dogmatic Freud, indecisive James, pedantic Pavlov, cocky Galton, dashing Boas – their (innate?) personalities were too disparate, their (acquired?) cultural backgrounds too diverse and their whiskers would have got tangled.

I suppose it is possible that they could have sorted the mess out at the beginning and avoided a century of dispute about nature and nurture. They could have granted Darwin, James and Galton the innateness of personality; granted De Vries the particulate nature of inheritance; granted Kraepelin, Freud and Lorenz a role for early experience in shaping the psyche; granted Piaget the importance of developmental stages; granted Pavlov and Watson the power of learning to reshape the adult mind; granted Boas and Durkheim the autonomous power of culture and society. All these things could be true at the same time, they could have said. Learning could not happen without an innate capacity to learn. Innateness could not be expressed without experience. The truth of each idea is not proof of the falsehood of another.

Possible, but not likely. Even if they had achieved this – for philosophers – superhuman feat, I cannot see them binding those who followed them to the treaty. Hostilities would have resumed soon enough between the partisans of different theories: it's in human nature. There seems to be something almost inevitable about dividing human psychology into nature and nurture. Perhaps, as Sarah Hrdy has suggested, the dichotomy is itself an instinct – in the genes. Instead of a stately progress towards enlightenment, the twentieth century became a collision of ideas, a hundred years war between the forces of nature and the forces of nurture. Anthropology was its Flanders, Harvard its Manassas, Russia its Russia. Remaining neutral was difficult; those who kept the respect of both sides, such as John Maynard Smith and Pat Bateson, found it hard going. Too many slipped into the false equation that to prove one proposition right was to prove another wrong – that success for nature could only mean defeat for nurture, or vice versa. Even as they repeated the platitude, 'Of course, it's both', many could not resist the temptation to see it in zero-sum terms, like a battle. I hope I have shown in this book how mistaken this is. I hope I have shown that the more you discover genes that influence behaviour, the more you find that they work through nurture, and the more you find that animals learn, the more you discover that learning works through genes.

Bizarrely, even the fiercest warriors of the hundred years war knew this. The following quotations are all from veterans of those wars. Can you tell which side they were on?

[I view] humans as dynamic, creative organisms for whom the opportunity to learn and to experience new environments amplifies the effect of the genotype on the phenotype.[2]

Each person is molded by an interaction of his environment, especially his cultural environment, with the genes that affect social behavior.[3]

Where on earth did the myth of the inevitability of genetic effects come from?[4]

If my genes don't like it, they can jump in the lake.[5]

In so far as any aspect of life can be said to be in the 'genes', our genes provide the capacity for both specificity – a lifeline relatively impervious to developmental and environmental buffering – and plasticity – the ability to respond appropriately to unpredictable environmental contingency.[6]

If we are programmed to be what we are, then these traits are ineluctable. We may, at best, channel them, but we cannot change them either by will, education or culture.[7]

An organism's genes, to the extent that they influence what the organism does, in its behavior, physiology, and morphology, are at the same time helping to construct an environment.[8]

I'm a reductionist and a geneticist. Memory is, in a sense, the sum of all memory genes.[9]

The quotes are from Thomas Bouchard, Edward Wilson, Richard Dawkins, Steven Pinker, Steven Rose, Stephen Gould, Richard Lewontin and Tim Tully. The first four would be considered extreme genetic determinists by the second four. Yet, in truth, each of these polemicists believes roughly the same thing. He believes that human nature comes from an interaction of nature with nurture. Only his opponent holds immoderate views. But his opponent is a straw man.

In the history of the nature–nurture debate, the truly great breakthroughs, the moments of startling enlightenment, were impossible to categorise as victories for either side. The experiments I have celebrated in this book – Lorenz's goslings, Harlow's monkeys, Mineka's toy snakes, Insel's voles, Zipursky's flies, Rankin's worms, Holt's tadpoles, Blanchard's brothers, Moffitt's children – in each and every case provide evidence of genes that work by reacting to experience. The Lorenzian gosling is genetically programmed to imprint on whatever the environment provides as a model parent. The Harlovian monkey is genetically inclined to prefer certain kinds of mother, but cannot develop properly without maternal love. The Minekan snake elicits an instinctive phobia, but only if paired with a fearful reaction from a model. The Inselar vole is programmed to fall in love, but only in response to certain experiences. The Zipurskian fly's eyes are equipped with genes that feel their way into the brain responding to the environment they find along the way. The Rankinian worms alter the expression of their genes in response to schooling. The Holtian tadpole has growth cones on the tips of its neurons that express genes in response to the world around them. The Blanchardian womb of a mother of many sons is made more likely, through her genes, to cause her next son to be gay. A Moffittian abused child is nurtured to antisocial behaviour, but only if equipped with a certain version of a gene. These are truly the experiments that show genes to be the epitomes of sensitivity, the means by which creatures can be flexible, the very servants of experience. Nature versus nurture is dead. Long live nature via nurture.

Biarritz, 1st April 1903. Left to right: Sigmund Freud, Franz Boaz, John Watson, William James, Jean Piaget, Lorenz, Emile Durkheim, Hugo De Vries, Pavlov, Darwin, Galton, Emil Kraepelin.

OUT

ACKNOWLEDGEMENTS

Great thanks to all the scientists who shared their astonishing nuggets from the genome with me, and to all those who cleared my head of nonsense and filled it with better ideas. Some people gave me long interviews, others responded to emails and some just contributed passing remarks. They all did so with unfettered generosity. They include Michael Bailey, Simon Baron-Cohen, Pat Bateson, Ray Blanchard, Dorret Boomsma, Tom Bouchard, John Burn, Ira Carmen, Sue Carter, Avshalom Caspi, Shirley Chan, Hollis Cline, Steve Cohen, Peter Corning, Leda Cosmides, Francis Crick, Tim Crow, Tony Curzon-Price, Richard Dawkins, Paromita Deb-Rinker, Mickey Diamond, Alan Dixson, Sean Eddy, Thalia Eley, Mike Fainzilber, James Flynn, Alex Gann, Mary-Jane Gething, David Goetze, Anthony Gottlieb, Jean-Pierre Hardelin, Judith Rich Harris, Scott Hawley, Andrew Holmes, Gabriel Horn, Sarah Hrdy, Josh Huang, Tim Hubbard, Tom Insel, Bill Irons, Lucia Jacobs, Randal Keynes, Jonathan Kingdon, Tom Kirkwood, Robert Krueger, Robb Krumlauf, Naida Loskutoff, Robin Lovell-Badge, Bobbi Low, Hugh Lytton, Zach Mainen, Nick Martin, Roger Masters, Brian McCabe, Robin McKie, Chris McManus, Michael Meaney, Drew Mendelsohn, David Micklos, Geoffrey Miller, Sue Mineka, Graeme Mitchison, Terrie

Moffitt, Bill Neaves, Randy Nesse, John Orbell, Svante Paabo, Steven Pinker, Robert Plomin, Malcolm Potts, Cathy Rankin, Mark Ridley, Giacomo Rizzolatti, Pemilla Roth, Joe Sambrook, Ken Schaffner, Nancy Segal, Phil Sharp, Richard Sherlock, Neil Smalheiser, Tim Specter, Robert Sprinkle, David Stern, David Stewart, Bruce Stillman, John Sulston, Ian Tattersall, Bronwyn Terrill, John Tooby, Patricia Tueting, Tim Tully, Eric Turkheimer, Ajit Varki, Richard Viken, Christopher Walsh, Jim Watson, Mary-Jane West-Eberhard, Jan Witkowski, Geoffrey Woods, Robert Wozniak, Richard Wrangham, Pat Wright, Robert Yolken and Larry Zipursky.

While writing the book, I was fortunate to spend some time in the intellectually exciting yet aesthetically peaceful surroundings of Cold Spring Harbor on Long Island. I am very grateful to all those who made our stay there such a pleasure, especially Jim and Liz Watson, Bruce and Grace Stillman and Jan and Fiona Witkowski. I am also especially grateful to my hosts at the Stowers Institute in Kansas City when the awful events of 11 September 2001 briefly stranded me there: especially Bill Neaves and Neil and Jean Patterson. Back home I thank all my colleagues at the International Centre for Life for their support and encouragement over the past two years, including Alastair Balls, Linda Conlon, Steve Cross and Teresa McDonald.

Several people generously read all or part of the book in draft and suggested vital changes: Richard Dawkins, Graeme Mitchison, Randy Nesse, Jim Watson, John Tooby, Anya Hurlbert.

My editors, Terry Karten and Christopher Potter, gave me plenty of rope, my agents, Felicity Bryan and Peter Ginsberg, did a fabulous job as ever, and both of my publishers turned the book around in record time.

The phrase 'nature via nurture' was first coined by David Lykken, who kindly allowed me to use it as the title of my book. Anya Hurlbert's advice and support – neuroscientific, literary and personal – was priceless throughout.

END NOTES

PROLOGUE

1. Book 1, Line 58.
2. *Observer*, 11 February 2001.
3. *San Francisco Chronicle*, 11 February 2001.
4. *New York Times*, 12 February 2001.
5. See http://web.fccj.org/~ethall/trivia/solvay.htm

CHAPTER I: THE PARAGON OF ANIMALS

1. Act 3, scene 4.
2. Keynes, R.D. (ed.). 1988. *Charles Darwin's Beagle Diary*. Cambridge University Press.
3. Ibid.
4. Keynes, R.D. 2001. *Annie's Box*. 4th Estate.
5. Quoted in Degler, C.N. 1991. *In Search of Human Nature*. Oxford University Press.
6. Quoted in Midgely, M. 1978. *Beast and Man*. Routledge.
7. Budiansky, S. 1998. *If a Lion Could Talk*. Weidenfeld & Nicolson.

8. *Buffon's Natural History* (abridged). 1792. London.
9. Bewick, T. 1807. *A General History of Quadrupeds*. Newcastle upon Tyne.
10. Morris, R. and Morris, D. 1966. *Men and Apes*. Hutchinson.
11. Goodall, J. 1990. *Through a Window*. Houghton Mifflin.
12. Ibid.
13. Rendell, L. and Whitehead, H. 2001. Culture in whales and dolphins. *Behavioural and Brain Sciences* 24:309–24.
14. Call, J. 2001. Chimpanzee social cognition. *Trends in Cognitive Science* 5:388–93.
15. Malik, K. 2001. *What Is It to Be Human?*. Institute of Ideas.
16. Darwin, C. 1871. *The Descent of Man*. John Murray.
17. Malik, K. 2001. *What Is It to Be Human?*. Institute of Ideas.
18. Midgley, M. 1978. *Beast and Man*. Routledge.
19. Zuk, M. 2002. *Sexual Selections*. University of California Press.
20. van Schaik, C.P. and Kappeler, P.M. 1997. Infanticide risk and the evolution of male–female association in primates. *Proceedings of the Royal Society* B:264:1687–94.
21. Wrangham, R.W., Jones, J.H., Laden, G., Pilbeam, D. and Conkin-Brittain, N. 1999. The Raw and the Stolen. Cooking and the ecology of human origins. *Current Anthropology* 40:567–94.
22. Ridley, M. 1996. *The Origins of Virtue*. Penguin.
23. Wrangham, R.W. and Peterson, D. 1997. *Demonic Males*. Bloomsbury.
24. Alan Dixson, email correspondence.
25. http://www.blockbonobofoundation.org.
26. Ebersberger, I., Metzier, D., Schwarz, C. and Paabo, S. 2002. Genome-wide comparison of DNA sequences between human and chimpanzees. *American Journal of Human Genetics* 70:1490–97.
27. Britten, R.J. 2002. Divergence between samples of chimpanzee and human DNA sequences is 5%, counting indels. *Proceedings of the National Academy of Sciences* 99:13633–5.
28. King, M.C. and Wilson, A.C. 1975. Evolution at two levels in humans and chimpanzees. *Science* 188:107–16.
29. Sibley, C.G. and Ahiquist, J.E. 1984. The phylogeny of the hominoid primates, as indicated by DNA-DNA hybridization. *Journal of Molecular Evolution* 20:2–15.
30. Johnson, M.E., Viggiano, L., Bailey, J.A., Abdul-Rauf, M., Goodwin, G., Rocchi, M. and Eichler, E.E. 2001. Positive selection of a gene during the emergence of humans and African apes. *Nature* 413:514–19.

31. Hayakawa, T., Satta, Y., Gagneux, P., Varki, A. and Takahata, N. 2001. Alu-mediated inactivation of the human CMP-N-acetyineuraminic acid hydroxylase gene. *Proceedings of the National Academy of Sciences* 98:11399–404.

32. Ajit Varki, interview. See also Chou, H.-H. *et al.* 1998. A mutation in human CMP-sialic acid hydroxylase occurred after the Homo-Pan divergence. *Proceedings of the National Academy of Sciences* 95:11751–6; Gagneux, P. and Varki, A. 2001. Genetic differences between humans and great apes. *Molecular Phylogenetics and Evolution* 18:2–13; Varki, A. 2001. Loss of N-glycolylneuraminic acid in humans: mechanisms, consequences, and implications for hominid evolution. *Yearbook of Physical Anthropology* 44:54–69

33. Hammer, C.J., Tyler, H.D., Loskutoff, N.M., Armstrong, D.L., Funk, D.J., Lindsey, B.R. and Simmons, L.G. 2001. Compromised development of calves (*Bos gaurus*) derived from in vitro-generated embryos and transferred interspecifically into domestic cattle (*Bos taurus*). *Theriogenology* 55:1447–55; Loskutoff, N., email correspondence.

34. There is some confusion over the terminology here. Some biologists use 'promoter' to mean the site where the RNA polymerase enzyme binds after being recruited by a transcription factor. Here I use it in the broader sense, to mean the entire regulatory sequence of the gene.

35. Belting, H.G., Shashikant, C.S. and Ruddle, F.H. 1998. Modification of expression and cis-regulation of Hoxc8 in the evolution of diverged axial morphology. *Proceedings of the National Academy of Sciences* 95:2355–60.

36. Cohn, M.J. and Tickle, C. 1999. Developmental basis of limblessness and axial patterning in snakes. *Nature* 399:474–9.

37. Ptashne, M. and Gann, A. 2002. *Genes and Signals.* Cold Spring Harbor Press; also Alex Gann, interviews.

38. Carroll, S.B. 2000. Endless forms: the evolution of gene regulation and morphological diversity. *Cell* 101:577–80.

39. Coppinger, R. and Coppinger, L. 2001. *Dogs: a Startling New Understanding of Canine Origin, Behavior and Evolution.* Scribner.

40. Semendeferi, K., Armstrong, E., Schleicher, A., Zilles, K., and van Hoesen, G.W. 1998. Limbic frontal cortex in hominoids: a comparative study of area 13. *American Journal of Physical Anthropology* 106:129–55.

41. Wrangham, R.W., Pilbeam, D. and Hare, B. (unpublished). Convergent paedomorphism in bonobos, domesticated animals and humans: the role of selection for reduced aggression.

42. Wrangham, R.W. and Pilbeam, D. 2001, in *All Apes Great and Small,* volume 1; *Chimpanzees, Bonobos, and Gorillas* (ed. Galdikas, B., Erickson, N.

and Sheeran, L.K.). Plenum; also Wrangham, R.W. Talk at Cold Spring Harbor, President's Council, May 2001.
43. Quoted in the *New York Times*, 24 September 2002.
44. Bond, J., Roberts, E., Mochida, G.H., Hampshire, D.J., Scott, S., Askham, J.M., Springell, K., Mahadevan, M., Crow, Y.J., Markham, A.F., Walsh, C.A. and Woods, C.G. 2002. ASPM is a major determinant of cerebral cortical size. *Nature Genetics* 32:316–20.

CHAPTER 2: A PLETHORA OF INSTINCTS

1. Spalding, D.A. 1873 Instinct: with original observations on young animals. *Macmillan's Magazine* 27:282–93.
2. Myers, G.E. 1986. *William James: His Life and Thought.* Yale University Press.
3. Bender, B. 1996. *The Descent of Love: Darwin and the Theory of Sexual Selection in American Fiction, 1871–1926.* University of Pennsylvania Press.
4. James, W. 1890. *The Principles of Psychology.* Henry Holt.
5. Myers, G.E. 1986. *William James: His Life and Thought.* Yale University Press.
6. Dawkins, R. 1986. *The Blind Watchmaker.* Norton.
7. Dennett, D. 1995. *Darwin's Dangerous Idea.* Penguin.
8. James, W. 1890. *The Principles of Psychology.* Henry Holt.
9. Insel, T.R. and Shapiro, L.E. 1992. Oxytocin receptor distribution reflects social organization in monogamous and polygamous voles. *Proceedings of the National Academy of Sciences* 89:5981–5.
10. Argiolas, A., Melis, M.R., Stancampiano, R. and Gessa, G.L. 1989. Penile erection and yawning induced by oxytocin and related peptides: structure-activity relationship. *Peptides* 10:559–63.
11. Insel, T.R. and Shapiro, L.E. 1992. Oxytocin receptor distribution reflects social organization in monogamous and polygamous voles. *Proceedings of the National Academy of Sciences* 89:5981–5.
12. Ferguson, J.N., Young, L.J., Hearn, E.F., Matzuk, M.M., Insel, T.R. and Winslow, J.T. 2000. Social amnesia in mice lacking the oxytocin gene. *Nature Genetics* 25:284–8.
13. Young, L.J., Wang, Z. and Insel, T.R. 1998. Neuroendocrine bases of monogamy. *Trends in Neurosciences* 21:71–5.

14. Insel, T.R., Winslow, J.T., Wang, Z. and Young, L.J. 1998. Oxytocin, vasopressin, and the neuroendocrine basis of pair bond formation. *Advances in Experimental and Medical Biology* 449:215–24.
15. Insel, T.R. and Young, L.J. 2001. The neurobiology of attachment. *Nature Reviews in Neuroscience* 2:129–36.
16. Wang, Z., Yu, G., Cascio, C., Liu, Y., Gingrich, B. and Insel, T.R. 1999. Dopamine D2 receptor-mediated regulation of partner preference in female prairie voles (*Microtus ochrogaster*): a mechanism for pair bonding? *Behavioral Neuroscience* 113:602–11.
17. Jankowiak, W.R. and Fisher, E.F. 1992. A cross-cultural perspective on romantic love. *Ethnology* 31:149–55.
18. Insel, T.R., Gingrich, B.S. and Young, L.J. 2001. Oxytocin: who needs it? *Progress in Brain Research* 133:59–66.
19. Bartels, A. and Zeki, S. 2000. The neural basis of romantic love. *Neuroreport* 11:3829–34.
20. Carter, C.S. 1998. Neuroendocrine perspectives on social attachment and love. *Psychoneuroendocrinology* 23:779–818.
21. Ridley, M. 1993. *The Red Queen*. Penguin.
22. Tinbergen, N. 1951. *The Study of Instinct.* Oxford University Press.
23. Ginsburg, B.E. 2001. Fellow travellers on the road to the genetics of behavior: mice, rats and dogs. Talk to the International Behavioural and Neural Genetics Society, 8–10 November 2001, San Diego.
24. Konner, M. 2001. *The Tangled Wing: Biological Constraints on the Human Spirit.* 2nd edition. W.H. Freeman.
25. Budiansky, S. 2000. *The Truth about Dogs*. Viking Penguin.
26. You can check out such a bull catalogue at www.genusplc.com.
27. Eibl-Eibesfeldt, I. 1989. *Human Ethology*. Aldine de Gruyter; Ekman, P. 1998. Afterword: Universality of emotional expression? A personal history of the dispute. In Darwin, C., *The Expression of the Emotions in Man and Animals* (new edition). Oxford University Press.
28. Buss, D.M. 1994. *The Evolution of Desire*. Basic Books.
29. Buss, D.M. 2000. *The Dangerous Passion*. Bloomsbury.
30. You can find this quoted almost anywhere on the Internet.
31. Diamond, M., 1965. A critical evaluation of the ontogeny of human sexual behavior. *Quarterly Review of Biology* 40:147–75.
32. Colapinto, J. 2000. *As Nature Made Him: the Boy Who Was Raised as a Girl.* HarperCollins.
33. Reiner, W.G. 1999. Assignment of sex in neonates with ambiguous

genitalia. *Current Opinion in Pediatrics* 11:363–5. Also article in *The Times* (London), 26 June 2001, by Lisa Melton: Ethics and gender.

34. Lutchmaya, S., Baron-Cohen, S. and Raggatt, P. In press. Foetal testosterone and eye contact in 12 month old human infants. *Infant Behaviour Development* (in press).

35. Connellan, J., Baron-Cohen, S., Wheelwright, S., Batki, A. and Ahluwalia, J. 2000. Sex differences in human neonatal social perception. *Infant Behavior and Development* 23:113–18.

36. Baron-Cohen, S. 2002. The extreme male brain theory of autism. *Trends in Cognitive Sciences* 6:248–54.

37. Baron-Cohen, S. 2002. Autism: deficits in folk psychology exist alongside superiority in folk physics. In *Understanding Other Minds* (ed. Baron-Cohen, S., Tager-Flusberg, H. and Cohen, D.J.), pp. 73–82; Baron-Cohen, S., Wheelwright, S., Skinner, R., Martin, J. and Clubley, E. 2001. The autism spectrum quotient: evidence from Asperger syndrome/high-functioning autism, males and females, scientists and mathematicians. *Journal of Autism and Developmental Disorders* 31:5–17.

38. Baron-Cohen, S., Interview.

39. Frith, C. and Frith, U. 2000. The physiological basis of theory of mind: functional neuroimaging studies. In *Understanding Other Minds* (ed. Baron-Cohen, S., Tager-Flusberg, H. and Cohen, D.J.), pp. 334–56.

40. Tooby, J. and Cosmides, L. 1992. The psychological foundations of culture. In *The Adapted Mind* (ed. Barkow, J.H., Cosmides, L. and Tooby, J.). Oxford University Press.

41. Pinker, S. 1994. *The Language Instinct.* HarperCollins.

42. Sharma, J., Angelucci, A. and Sur, M. 2000. Induction of visual orientation modules in auditory cortex. *Nature* 404:841–7.

43. Finlay, B.L., Darlington, R.B. and Nicastro, N. 2001. Developmental structure in brain evolution. *Behavioral and Brain Sciences* 24:263–308.

44. Barton, R.A. and Harvey, P.H. 2000. Mosaic evolution of brain structure in mammals. *Nature* 405:1055–8.

45. Fodor, J. 2001. *The Mind Doesn't Work That Way.* MIT Press.

46. Pinker, S. 1997. *How the Mind Works.* Norton.

47. Lee, D. 1987. Introduction to Plato. *The Republic.* Penguin.

48. Neville-Sington, P. and Sington, D. 1993. *Paradise Dreamed: How Utopian Thinkers Have Changed the World.* Bloomsbury.

CHAPTER 3: A CONVENIENT JINGLE

1. Conversation with the author, Montreal, 2002.
2. Galton, F. 1869. *Hereditary Genius.*
3. Candolle, A. de. 1872. *Histoire des sciences et des savants depuis deux siècles.*
4. Galton, F. 1874. *English Men of Science: Their Nature and Nurture.*
5. *The Tempest*, Act 4, scene 1.
6. The text of Mulcaster's 'Positions' can be found at http://www.ucs.mun.ca/~wbarker/positions-txt.html
7. *A Midsummer Night's Dream*, Act 3, scene 2.
8. Galton, F. 1875. The history of twins, as a criterion of the relative powers of nature and nurture. *Fraser's Magazine* 12:566–76.
9. Gilham, N. 2001. *A Life of Sir Francis Galton: from African Exploration to the Birth of Eugenics.* Oxford University Press.
10. Ridley, M. 1999. *Genome.* Fourth Estate.
11. Lifton, R.J. 1986. *The Nazi Doctors.* Basic Books.
12. Wright, W. 1999. *Born That Way.* Routledge.
13. For an excellent summary of the ins and outs of twins, see Segal, N. 1999. *Entwined Lives.* Dutton. Incidentally, it is increasingly unfashionable to use the terms 'identical' and 'fraternal', researchers preferring the more precise 'monozgotic' and 'dizygotic'. But this is a popular book, so I stick to the popular terms.
14. For a general review of behaviour genetics, see: Plomin, R., DeFries, J.C., Craig, I.W. and McGuffin, P. 2002. *Behavioral Genetics in the Postgenomic Era.* American Psychological Association.
15. Wright, W. 1999. *Born That Way.* Routledge.
16. Farber, S.L. 1981. *Identical Twins Reared Apart: A Reanalysis.* Basic Books.
17. There is evidence that fraternal twins are actually more similar genetically than siblings, because although they come from different sperm, they often come from the same maternal oocyte, two of whose pronuclei develop into eggs. This, however, only makes it more remarkable that they prove so different from each other in personality compared with identical twins.
18. Bouchard, T.J., McGue, M., Lykken, D. and Tellegen, A. 1999. Intrinsic and extrinsic religiousness: genetics and environmental influences and personality correlates. *Twin Research* 2:88–98; Kirk, K.M., Eaves, L.J. and Martin, N. 1999. Self-transcendence as a measure of spirituality in a sample of older Australian twins. *Twin Research* 2:81–7.
19. Nelkin, D. and Lindee, M.S. 1996. *The DNA Mystique.* W.H. Freeman.

20. Pioneer Fund website.

21. Quoted in Wright, W. 1999. *Born That Way*. Routledge.

22. Pinker, S. 2002. *The Blank Slate*. Penguin.

23. Eley, T.C., Lichtenstein, P. and Stevenson, J. 1999. Sex differences in the etiology of aggressive and nonaggressive antisocial behavior: results from two twin studies. *Child Development* 70:155–68.

24. Mischel, W. 1981. *Introduction to Personality*. Holt, Rinehart and Winston.

25. Thomas Bouchard, interview.

26. Clark, W.R. and Grunstein, M. 2000. *Are We Hard-wired? The Role of Genes in Human Behavior*. Oxford University Press.

27. Bouchard, T.J. Jr. 1999. Genes, environment and personality, pp. 98–103 in *The Nature–Nurture Debate* (ed. Ceci, S.J. and Williams, W.M.). Blackwell.

28. Krueger, R. 2001. Talk to the 10th International Congress of Twin Studies, London, 4–7 July 2001.

29. Grilo, C.M. and Pogue-Geile, M.F. 1991. The nature of environmental influences on weight and obesity. *Psychological Bulletin* 110:520–37.

30. Randolph Nesse, email. See also Srijan, S., Nesse, R.M., Stoltenberg, S.F., Li, S., Gleiberman, L., Chakravarti, A., Weder, A.B. and Burmeister, M. 2002. A BDNF coding variant is associated with the NEO personality inventory domain neuroticism, a risk factor for depression. *Neuropsychopharmacology* (in press). Originally published 27 August 2002 at http://www.acnp.org/citations/Npp082902374

31. Bouchard, T.J. Jr, Lykken, D.T., McGue, M., Segal, N.L. and Tellegen, A. 1990. Sources of human psychological differences: the Minnesota Study of Twins Reared Apart. *Science* 250:223–8.

32. Eaves, L., D'Onofrio, B. and Russell, R. 1999. Transmission of religion and attitudes. *Twin Research* 2:59–61.

33. Tully, T., interview.

34. Turkheimer, E. 1998. Heritability and biological explanation. *Psychology Review* 105:782–91.

35. Zach Mainen, interview.

36. Jensen, A. 1969. How much can we boost IQ and scholastic achievement? *Harvard Educational Review* 39:1–123.

37. Herrnstein, R.J. and Murray, C. 1994. *The Bell Curve: Intelligence and Class Structure in American Life*. Free Press.

38. Posthuma, D., Neale, M.C., Boomsma, D.I. and de Geus, E.J. 2001. Are smarter brains running faster? Heritability of alpha peak frequency, IQ, and their interrelation. *Behavior Genetics* 31:567–79.

39. Thompson, P.M., Cannon, T.D., Narr, K.L., van Erp, T., Poutanen, V.-P., Huttunen, M., Lohnqvist, J., Standertskjold-Nordenstam, C.-G., Kaprio, J., Khaledy, M., Dail, R., Zoumalan, C.I. and Toga, A.W. 2001. Genetic influences on brain structure. *Nature Neuroscience* 4:1253–8; Posthuma, D., de Geus, E.J., Baare, W.F., Hulshoff Pol, H.E., Kahn, R.S. and Boomsma, D.I. 2002. The association between brain volume and intelligence is of genetic origin. *Nature Neuroscience* 5:83–4.
40. Turkheimer, E., Haley, A., D'Onofrio, B., Waldron, M., Emery, R.E. and Gottesman, I. 2001. Socioeconomic status modifies heritability of intelligence in impoverished children. Paper at the 2001 meeting of the Behavior Genetics Association annual meeting, Cambridge, July 2001.
41. McGue, M., Bouchard, T.J. Jr, Iacono, W.G. and Lykken, D.T. 1993. Behavior genetics of cognitive ability: a life-span perspective. In *Nature Nurture and Psychology* (ed. Plomin, R. and McClearn, G.E.), American Psychological Association; also McClearn, G.E. *et al.* 1997. Substantial genetic influence on cognitive abilities in twins 80+ years old. *Science* 276:1560–3.
42. Eley, T., interview.
43. Dickens, W.T. and Flynn, J.R. 2001. Heritability estimates versus large environmental effects: the IQ paradox resolved. *Psychological Review* 108:346–69.
44. Williams, A.G., Rayson, M.P., Jubb, M., World, M., Woods, D.R., Hayward, M., Martin J., Humphries, S.E. and Montgomery, H.E. 2000. The ACE gene and muscle performance. *Nature* 403:614.
45. Ridley, M. 1993. *The Red Queen.* Penguin.
46. Radcliffe-Richards, J. 2000. *Human Nature after Darwin.* Routledge.
47. Flynn, J.R. (unpublished). The history of the American mind in the 20th century: a scenario to explain IQ gains over time and a case for the irrelevance of g.
48. For those curious about Galton's unpublished novel, a fuller précis is given in Nicholas Gilham's biography of Galton, cited above.

CHAPTER 4: THE MADNESS OF CAUSES

1. James, W. 1890. *Principles of Psychology.*
2. Quoted in Shorter, E. 1997. *A History of Psychiatry.* John Wiley & Sons.

3. Fromm-Reichmann, F. 1948. Notes on the development of treatment of schizophrenics by psychoanalytic psychotherapy. *Psychiatry* 11:263–73.
4. Pollak, R. 1997. *The Creation of Dr B: a Biography of Bruno Bettelheim*. Simon & Schuster.
5. Folstein. S.E. and Mankoski, R.E. 2000. Chromosome 7q: where autism meets language disorder? *American Journal of Human Genetics* 67:278–81.
6. James, O. 2002. *They F*** You Up: How to Survive Family Life*. Bloomsbury.
7. The psychiatrist and writer Randolph Nesse calls this the central error of psychiatric research.
8. Cited in Torrey, E.F. 1988. *Surviving Schizophrenia: a Family Manual*. Harper and Row.
9. Shorter, E. 1997. *A History of Psychiatry*. John Wiley & Sons.
10. Wahlberg, K.E., Wynne, L.C., Oja, H. *et al.* 1997. Gene-environment interaction in vulnerability to schizophrenia: findings from the Finnish adoptive family study in schizophrenia. *American Journal of Psychiatry* 154:355–62.
11. Kety, S.S. and Ingraham, L.J. 1992. Genetic transmission and improved diagnosis of schizophrenia from pedigrees of adoptees. *Journal of Psychiatric Research* 26:247–55.
12. Tsuang, M., Stone, W.S. and Faraone, S.V. 2001. Genes, environment and schizophrenia. *British Journal of Psychiatry* 178 (supplement 40):s18–s24.
13. Sherrington, R., Brynjolfsson, J., Petursson, H. *et al.* 1988. Localization of a susceptibility locus for schizophrenia of chromosome 5. *Nature* 336:164–7; Bassett, A.S., McGillvray, B.C., Jones, B.D. *et al.* 1988. Partial trisomy of chromosome 5 cosegregating with schizophrenia. *Lancet* 1988:799–801.
14. Levinson, D.F. and Mowry, B.J. 1999. Genetics of schizophrenia. In *Genetic Influences on Neural and Behavioral Functions* (ed. Pfaff, D.W., Joh, T. and Maxson, S.C.), pp. 47–82. CRC Press, Boca Raton
15. Mirnics, K., Middleton, F.A., Lewis, D.A. and Levitt, P. 2001. Analysis of complex brain disorders with gene expression microarrays: schizophrenia as a disease of the synapse. *Trends in Neurosciences* 24:479–86.
16. Tsuang, M., Stone, W.S. and Faraone, S.V. 2001. Genes, environment and schizophrenia. *British Journal of Psychiatry* 178 (supplement 40): s18–s24.
17. Mednick. S.A., Machon, R.A., Huttunen, M.O., Bonett, D. 1988. Adult schizophrenia following prenatal exposure to an influenza epidemic. *Archives of General Psychiatry* 45:189–92; Munk-Jorgensen, P. and Ewald, H. 2001. Epidemiology in neurobiological research: exemplified by the influenza-schizophrenia theory. *British Journal of Psychiatry* 178 (supplement 40):s30–s32.

18. Davis, J.O., Phelps, J.A. and Bracha, H.S. 1999. Prenatal development of monozygotic twins and concordance for schizophrenia. In *The Nature–Nuture Debate* (ed. Ceci, S.J. and Williams, W.W.). Blackwell.
19. Tsuang, M., Stone, W.S. and Faraone, S.V. 2001. Genes, environment and schizophrenia. *British Journal of Psychiatry* 178 (supplement 40): s18–s24.
20. Deb-Rinker, P., Klempan, T.A., O'Reilly, R.L., Torrey, E.F. and Singh, S.M. 1999. Molecular characterization of a MSRV-like sequence identified by RDA from monozygotic twin pairs discordant for schizophrenia. *Genomics* 61:133–44.
21. Karlsson, H., Bachmann, S., Schroder, J., McArthur, J., Torrey, E.F. and Yolken, R.H. 2001. Retroviral RNA identified in the cebrebrospinal fluids and brains of individuals with schizophrenia. *Proceedings of the National Academy of Sciences* 98:4634–9.
22. Impagatiello, F., Guidotti, A.R., Pesold, C. *et al.* 1998. A decrease of reelin expression as a putative vulnerability factor in schizophrenia. *Proceedings of the National Academy of Sciences* 95:15718–23.
23. Fatemi, S.H., Emamian, E.S., Kist, D., Sidwell, R.W., Nakajima, K., Akhter, P., Shier, A., Sheikh, S. and Bailey, K. 1999. Defective corticogenesis and reduction in reelin immunoreactivity in cortex and hippocampus of prenatally infected neonatal mice. *Molecular Psychiatry* 4:145–54.
24. Fatemi, S.H. 2001. Reelin mutations in mouse and man: from reeler mouse to schizophrenia, mood disorders, autism and lissencephaly. *Molecular Psychiatry* 6:129–33.
25. Hong, S.E., Shugart, Y.Y., Huang, D.T., Shahwan, S.A., Grant, P.E., Hourihane, J.O., Martin, N.D. and Walsh, C.A. 2000. Autosomal recessive lissencephaly with cerebellar hypoplasia is associated with human RELN mutations. *Nature Genetics* 26:93–6.
26. Cannon, M., Caspi, A., Moffitt, T.E., Harrington, H., Taylor, A., Murray, R.M. and Poulton, R. 2002. Evidence for early-childhood, pan-developmental impairment specific to schizophreniform disorder: results from a longitudinal birth cohort. *Archives of General Psychiatry* 59:449–56.
27. Weinberger, D.R. 1987. Implications of normal brain development for the pathogenesis of schizophrenia. *Archives of General Psychiatry* 44:660–9. Weinberger, D.R. 1995. From neuropathology to neurodevelopment. *Lancet* 26:552–7.
28. Mirnics, K., Middleton, F.A., Lewis, D.A. and Levitt, P. 2001. Analysis of complex brain disorders with gene expression microarrays: schizophrenia as a disease of the synapse. *Trends in Neurosciences* 24:479–86.

29. Horrobin, D. 2001. *The Madness of Adam and Eve*. Bantam.
30. Peet, M., Glen, I. and Horrobin, D. 1999. *Phospholipid Spectrum Disorder in Psychiatry*. Marius Press.
31. Jablensky, A., Sartorius, N., Ernberg, G., Anker, M., Korten, A., Cooper, J.E., Day, R. and Bertelson, A. 1992. Schizophrenia: manifestations, incidence and course in different cultures. A World Health Organisation Ten Country Study, *Psychological Medicine Supplement* 20:1–97.
32. Quoted in Horrobin, D. 2001. *The Madness of Adam and Eve*. Bantam.
33. Stevens, A. and Price, J. 2000. *Prophets, Cults and Madness*. Duckworth, London.
34. Simonton, D.K. 2002. *The Origins of Genius*. Oxford University Press.
35. Nasar, S. 1998. *A Beautiful Mind: a biography of John Forbes Nash Jr.* Faber & Faber, London.

CHAPTER 5: GENES IN THE FOURTH DIMENSION

1. Dawkins, 1981. See http://www.world-of-dawkins.com/Dawkins/Work/Reviews/1985-01-24notinourgenes.htm.
2. Singer, D.G. and Revenson, T.A. 1996. *A Piaget Primer: How a Child Thinks* (2nd edition). Plume.
3. Lehrman, D.S. 1953. A critique of Konrad Lorenz's theory of instinctive behavior. *Quarterly Review of Biology* 28:337–63.
4. Tinbergen, N. 1963. On the aims and methods of ethology. *Zeitschrift für Tierpsychologie* 20:410–33.
5. Schaffner, K.F. 1998. Genes, behavior and developmental emergentism: one process, indivisible? *Philosophy of Science* 65:209–52.
6. West-Eberhard, M.J. 1998. Evolution in the light of cell biology, and vice versa. *Proceedings of the National Academy of Sciences* 95:8417–19.
7. For example. Oyama, S. 2000. *Evolution's Eye*. Duke University Press.
8. Greenspan, R.J. 1995. Understanding the genetic construction of behavior. *Scientific American*, April: 72–8.
9. Waddington, C.H. 1940. *Organisers and Genes*. Cambridge University Press.
10. Ariew, A. 1999. Innateness is canalization: in defense of a developmental account of innateness. In Hardcastle, V. (ed.) *Biology meets Psychology: Conjectures, Connections, Constraints*. MIT Press.

11. Bateson, P. and Martin, P. 1999. *Design for a Life: How Behaviour Develops*. Jonathan Cape.
12. See the review of 'Not in Our Genes' by Richard Dawkins, in *New Scientist*, 24 January 1985. Available online at http://www.world-of-dawkins.com/Dawkins/Work/Reviews/1985-01-24notinourgenes.htm.
13. Zhang, X. and Firestein, S. 2002. The olfactory receptor gene superfamily of the mouse. *Nature Neuroscience* 5:124–33.
14. Gogos, J.A., Osborne, J., Nemes, A., Mendelson, M. and Axel, R. 2000. Genetic ablation and restoration of the olfactory topographic map. *Cell* 103:609–20.
15. Wang, F., Nemes, A., Mendelsohn, M. and Axel, R. 1998. Odorant receptors govern the formation of a precise topographic map. *Cell* 93:47–60.
16. Holt, C. Lecture to Society for Neurosciences meeting, San Diego, November 2001; Campbell, D.S. and Holt, C.E. 2001. Chemotropic responses of retinal growth cones mediated by rapid local protein synthesis and degradation. *Neuron* 32:1013–26.
17. Tessier-Lavigne, M. and Goodman, C.S. 1996. The molecular biology of axon guidance. *Science* 274:1123–33; Yu, T.W. and Bargmann, C.I. 2001. Dynamic regulation of axon guidance. *Nature Neuroscience* 4 (Supplement): 1169–76.
18. Richards, L.J. 2002. Surrounded by Slit – how forebrain commissural axons can be led astray. *Neuron* 33:153–5.
19. Marillat, V., Cases, O., Nguyen-Ba-Charvel, K.T., Tessier-Lavigne, M., Sotelo, C. and Chedotal, A. 2002. Spatiotemporal expression patterns of slit and robo genes in the rat brain. *Journal of Comparative Neurology* 442:130–55; Dickson, B.J., Cline, H., Polleux, F. and Ghosh, A. 2001. Making connections: axon guidance and neural plasticity. *Embo Reports* 2:182–6.
20. Soussi-Yanicostas, N., Faivre-Sarrailh, C., Hardelin, J.-P., Levilliers, J., Rougon, G. and Petit, C. 1998. Anosmin-1 underlying the X chromosome-linked Kallman syndrome is an adhesion molecule that can modulate neurite growth in a cell-type specific manner. *Journal of Cell Science* 111:2953–65.
21. Hardelin, J.-P. 2001. Kallmann syndrome: towards molecular pathogenesis. *Journal of Molecular Endocrinology* 179:75–81.
22. Oliveira, L.M., Seminara, S.B., Beranova, M., Hayes, F.J., Valkenburgh, S.B., Schiphani, E., Costa, E.M., Latronico, A.C., Crowley, W.F., Vallejo, M. 2001. The importance of autosomal genes in Kallmann syndrome: genotype–phenotype correlations and neuroendocrine characteristics. *Journal of Clinical Endocrinology and Metabolism* 86:1532–8.

23. Dawkins, R. 1982. *The Extended Phenotype*. Oxford University Press.
24. Braitenburg, V. 1967. Patterns of projection in the visual system of the fly. I. Retina-lamina projections. *Experimental Brain Research* 3:271–98.
25. Lee, C.H., Herman, T., Clandinin, T.R., Lee, R. and Zipursky, S.L. 2001. N-cadherin regulates target specificity in the Drosophila visual system. *Neuron* 30:437–50; Clandinin, T.R., Lee, C.H., Herman, T., Lee, R.C., Yang, A.Y., Ovasapyan, S. and Zipursky, S.L. 2001. *Drosophila* LAR regulates R1-R6 and R7 target specificity in the visual system. *Neuron* 33:237–48. Also Zipursky, S.L., interview with the author, and talk to Society for Neuroscience, San Diego, November 2001.
26. Modrek, B. and Lee, C. 2002. A genomic view of alternative splicing. *Nature Genetics* 30:13–19.
27. Schmucker, D., Clemens, J.C., Shu, H., Worby, C.A., Xiao, J., Muda, M., Dixon, J.E. and Zipursky, S.L. 2000. Drosophila Dscam is an axon guidance receptor exhibiting extraordinary molecular diversity. *Cell* 101:671–84.
28. Serafini, T. 1999. Finding a partner in a crowd: neuronal diversity and synaptogenesis. *Cell* 98:133–6.
29. Wang, X., Su, H. and Bradley, A. 2002. Molecular mechanisms governing Pcdh-gamma gene expression: evidence for a multiple promoter and cis-alternative splicing model. *Genes and Development* 16:1890–905.
30. Wu, Q., and Maniatis, T. 1999. A striking organization of a large family of human neural cadherin-like cell adhesion genes. *Cell* 97:779–90; Tasic, B., Nabholz, C.E., Baldwin, K.K., Kim, Y., Rueckert, E.H., Ribich, S.A., Cramer, P., Wu, Q., Axel, R. and Maniatis, T. 2002. Promoter choice determines splice site selection in Protocadherin alpha and gamma Pre-mRNA splicing. *Molecular Cell* 10:21–33.
31. Specter, M. 2002. Rethinking the brain. In *Best American Science Writing 2002* (ed. M. Ridley). HarperCollins.
32. H. Cline, interview.
33. Gomez, M., De Castro, E., Guarin, E., Sasakura, H., Kuhara, A., Mori, I., Bartfai, T., Bargmann, C.I. and Nef, P. 2001. Ca2+ signalling via the neuronal calcium sensor-1 gene regulates associative learning and memory in C. elegans. *Neuron* 30:241–8.
34. Rankin, C., Rose, J. and Norman, K. 2001. The use of reporter genes to study the effects of experience on the anatomy of an identified synapse in the nematode C. elegans. Paper delivered at the IBANGS conference, San Diego, November 2001.

35. Harlow, H. and Harlow, M. 1962. Social deprivation in monkeys. *Scientific American* 207:136–46.
36. Meaney, M.J. 2001. Maternal care, gene expression and the transmission of individual differences in stress reactivity across generations. *Annual Reviews of Neuroscience* 24:1161–82.
37. Champagne, F., Diorio, J., Sharma, S. and Meaney, M.J. 2001. Naturally occurring variations in maternal behavior in the rat are associated with differences in estrogen-inducible central oxytocin receptors. *Proceedings of the National Academy of Sciences* 98:12736–41
38. Darlene D. Francis, Kathleen Szegda, Gregory Campbell, W. David Martin, Thomas R. Insel (unpublished). Epigenetic Sources of Behavioral Differences: Mother Nature Meets Mother Nurture.
39. Huxley, A. 1932. *Brave New World.* Chatto & Windus.

CHAPTER 6: FORMATIVE YEARS

1. *Paradise Regained* (1671), Book 4.
2. Quoted in Nisbett, A. 1976. *Konrad Lorenz.* Dent.
3. Nisbett, A. 1976. *Konrad Lorenz.* Dent.
4. Spalding, D.A. 1873. Instinct: with original observations on young animals. *Macmillan's Magazine* 27:282–93.
5. Bateson, P. 2000. What must be known in order to understand imprinting? in *The Evolution of Cognition* (ed. Heyes, C. and Huber, L.). MIT Press.
6. Gottlieb, G. 1997. *Synthesizing Nature–Nurture: Prenatal Roots of Instinctive Behavior.* Lawrence Erlbaum Associates.
7. Barker, D.J., Winter, P.D., Osmond, C., Margetts, B. and Simmonds, S.J. 1989. Weight in infancy and death from ischaemic heart disease. *Lancet* 8663:577–80.
8. Eriksson, J.G., Forsen, T., Tuomilehto, J., Osmond, C. and Barker, D.J. 2001. Early growth and coronary heart disease in later life: longitudinal study. *British Medical Journal* 322:949–53.
9. Bateson, P. 2001. Fetal experience and good adult design. *International Journal of Epidemiology* 30:928–34.
10. Manning, J., Martin, S., Trivers, R. and Soler, M. 2002. 2nd to 4th digit ratio and offspring sex ratio. *Journal of Theoretical Biology* 217:93.
11. Manning. J.T. and Bundred, P.E. 2000. The ratio of 2nd to 4th digit

length: a new predictor of disease predisposition? *Medical Hypotheses* 54:855–7; Manning, J.T., Baron-Cohen, S., Wheelwright, S. and Sanders, G. 2001. The 2nd to 4th digit ratio and autism. *Developmental Medicine and Child Neurology* 43:160–4.

12. Bischof, H.J., Geissler, E. and Rollenhagen, A. 2002. Limitations of the sensitive period of sexual imprinting: neuroanatomical and behavioral experiments in the zebra finch (*Taeniopygia guttata*). *Behavioral Brain Research* 133:317–22.

13. Burr, C. 1996. *A Separate Creation: How Biology Makes Us Gay*. Bantam Press.

14. Bailey, M., interview.

15. Symons, D. 1979. *Evolution of Human Sexuality*. Oxford University Press.

16. Blanchard, R. 2001. Fraternal birth order and the maternal immune hypothesis of male homosexuality. *Hormones and Behavior* 40:105–14.

17. Cantor, J.M., Blanchard, R., Paterson, A.D. and Bogaert, A.F. 2002. How may gay men owe their sexual orientation to fraternal birth order? *Archives of Sexual Behavior* 31:63–71.

18. Blanchard, R. and Ellis, L. 2001. Birth weight, sexual orientation and the sex of preceding siblings. *Journal of Biosocial Science* 33:451–67.

19. Blanchard, R., Zucker, K.J., Cavacas, A., Allin, S., Bradley, S.J. and Schachter, D.C. 2002. Fraternal birth order and birth weight in probably prehomosexual feminine boys. *Hormones and Behavior* 41:321–7.

20. Blanchard, R., Zucker, K.J., Cavacas, A., Allin, S., Bradley, S.J. and Schachter, D.C. 2002. Fraternal birth order and birth weight in probably prehomosexual feminine boys. *Hormones and Behavior* 41:321–7.

21. Harvey, R.J., McCabe, B.J., Solomonia, R.O., Horn, G. and Darlison, M.G. 1998. Expression of GABAa receptor gamma4 subunit gene: anatomical distribution of the corresponding mRNA in the domestic chick forebrain and the effect of imprinting training. *European Journal of Neuroscience* 10:3024–8.

22. Nedivi, E. 1999. Molecular analysis of developmental plasticity in neocortex. *Journal of Neurobiology* 41:135–47.

23. Huang, Z.J., Kirkwood, A., Pizzorusso, T., Porciatti, V., Morales, B., Bear, M.F., Maffei, L. and Tonegawa, S. 1999. BDNF regulates the maturation of inhibition and the critical period of plasticity in mouse visual cortex. *Cell* 98:739–55.

24. Fagiolini, M. and Hensch, T.K. 2000. Inhibitory threshold for critical-period activation in primary visual cortex. *Nature* 404:183–6.

25. Huang, J., interview.

26. Kegl, J., Senghas, A. and Coppola, M. 1999. Creation through contact: Sign language emergence and sign language change in Nicaragua. In *Comparative Grammatical Change: The Intersection of Language Acquisition, Creole Genesis, and Diachronic Syntax* (ed. DeGraff, M.). MIT Press; Bickerton, D. 1990. *Language and Species.* University of Chicago Press.

27. http://www.ling.lancs.ac.uk/monkey/ihe/linguistics/LECTURE4/4victor.htm. Newton, M. 2002. *Savage Girls and Wild Boys: A History of Feral Children.* Faber & Faber.

28. http://www.ling.lancs.ac.uk/monkey/ihe/linguistics/LECTURE4/4kaspar.htm.

29. Rymer, R. 1994. *Genie: a Scientific Tragedy.* Penguin.

30. Westermarck, E. 1891. *A History of Human Marriage.* Macmillan.

31. Wolf, A.P. 1995. *Sexual Attraction and Childhood Association: a Chinese Brief for Edward Westermarck.* Stanford University Press.

32. Shepher, J. 1971. Mate selection among second-generation kibbutz adolescents: incest avoidance and negative imprinting. *Archives of Sexual Behavior* 1:293–307.

33. Walter, A. 1997. The evolutionary psychology of mate selection in Morocco – a multivariate analysis. *Human Nature* 8:113–37.

34. Price, J.S. 1995. The Westermarck trap: a possible factor in the creation of Frankenstein. *Ethology and Sociobiology* 16:349–53.

35. Thornhill. N.W. 1991. An evolutionary analysis of rules regulating human inbreeding and marriage. *Behavioral and Brain Services* 14:247–60.

36. Greenber, M. and Littlewood, R. 1995. Post-adoption incest and phenotypic matching: experience, personal meanings and biosocial implications. *British Journal of Medical Psychology* 68:29–44.

37. Bevc, I. and Silverman, I. 1993. Early proximity and intimacy between siblings and incestuous behavior – a test of the Westermarck theory. *Ethology and Sociobiology* 14:171–81.

38. Deichmann, U. 1996. *Biologists under Hitler.* Harvard University Press.

39. Nisbett, A. 1976. *Konrad Lorenz*, Dent.

CHAPTER 7: LEARNING LESSONS

1. Turgenev, I. 1861/1975. *Fathers and Sons*. Penguin.
2. Todes, D.P. 1997. Pavlov's physiology factory. *Isis* 88:205–46.
3. Kimble, G.A. 1993. Evolution of the nature–nurture issue in the history of psychology. In *Nature Nurture and Psychology* (ed. Plomin, R. and McClearn, G.E.), American Psychological Association.
4. Frolov, Y.P. 1938. *Pavlov and His School.* Kegan Paul, Trench, Trubner & Co.
5. Waelti, P., Dickinson, A. and Schultz, W. 2001. Dopamine responses comply with basic assumptions of formal learning theory. *Nature* 412:43–8.
6. Watson, J.B. 1924. *Behaviorism.* W.W. Norton, New York.
7. Dubnau, J., Grady, L., Kitamoto, T. and Tully, T. 2001. Disruption of neurotransmission in Drosophila mushroom body blocks retrieval but not acquisition of memory. *Nature* 411:476–80.
8. Tully, T., interview.
9. Husi, H. and Grant, S.G.N. 2001. Proteomics of the nervous system. *Trends in Neurosciences* 24:259–66.
10. Watson, J.B. 1913. Psychology as the behaviorist views it. *Psychological Review* 20:158–77.
11. Rilling, M. 2000. John Watson's paradoxical struggle to explain Freud. *American Psychologist* 55:301–12.
12. Watson, J.B. and Rayner, R. 1920. Conditioned emotional reactions. *Journal of Experimental Psychology* 3:1–14.
13. Watson, J.B. 1924. *Behaviorism.* W.W. Norton.
14. Figes, O. 1996. *A People's Tragedy.* Jonathan Cape.
15. Frolov, Y.P. 1938. *Pavlov and His School.* Kegan Paul, Trench, Trubner & Co.
16. Figes, O. 1996. *A People's Tragedy.* Jonathan Cape.
17. All quotations from or about Lysenko are from Joravsky, D. 1986. *The Lysenko Affair.* University of Chicago Press.
18. Ibid.
19. Ibid.
20. Gould, S.J. 1978. *Ever Since Darwin.* Burnett Books.
21. Pinker, S. 2002. *The Blank Slate.* Penguin.
22. Blum, D. 2002. *Love at Goon Park.* Perseus Publishing.
23. Harlow, H.F. 1958. The nature of love. *American Psychologist* 13:673–85.

24. For a review of Mineka's work, see Ohman, A. and Mineka, S. 2001. Fears, phobias and preparedness: toward an evolved module of fear and fear learning. *Psychological Review* 108:483–522.
25. Fredrikson, M., Annas, P. and Wik, G. 1997. Parental history, aversive exposure and the development of snake and spider phobia in women. *Behavior Research Therapy* 35:23–8.
26. Ledoux, J. 2002. *Synaptic Self: How Our Brains Become Who We Are.* Viking.
27. Ohman, A. and Mineka, S. 2001, Fears, phobias and preparedness: toward an evolved module of fear and fear learning. *Psychological Review* 108:483–522.
28. Kendler, K.S., Jacobson, K.C., Myers, J. and Prescott, C.A. 2002. Sex differences in genetic and environmental risk factors for irrational fears and phobias. *Psychological Medicine* 32:209–17.
29. Hebb, D.O. 1949. *The Organization of Behavior: A Neuropsychological Theory.* Wiley.
30. Elman, J., Bates, E.A., Johnson, M.H., Karmiloff-Smith, A., Parisi, D. and Plunkett, K. 1996. *Rethinking Innateness.* MIT Press.
31. Ibid.
32. Fodor, J. 2001. *The Mind Doesn't Work That Way.* MIT Press.
33. Pinker, S. 2002. *The Blank Slate.* Penguin.
34. Skinner, B.F. 1948/1976. *Walden Two.* Prentice Hall.
35. See www.loshorcones.org.mx.

CHAPTER 8: CONUNDRUMS OF CULTURE

1. *Essay on Human Understanding*, 1692. Which only goes to show that Locke was not the blind blank-slater he has often been made out to be.
2. Kuper, A. 1999. *Culture: the Anthropologists' Account.* Harvard University Press.
3. Muller-White, L. 1998. *Franz Boas among the Inuit of Baffin Island, 1883–1884: Letters and Journals.* University of Toronto Press.
4. Quoted in Degler, C.N. 1991. *In Search of Human Nature.* Oxford University Press.
5. Ibid.

6. See *New York Times*, 8 October 2002, p. F3. Also: Sparks, C.S. and Jantz, R.L. 2002. A reassessment of human cranial plasticity: Boas revisited. *Proceedings of the National Academy of Sciences*. 8 October. 2002.
7. Freeman, D. 1999. *The Fateful Hoaxing of Margaret Mead: a Historical Analysis of Her Samoan Research*. Westview Press.
8. Durkheim, E. 1895. *The Rules of the Sociological Method*. (1962 edition). Free Press.
9. Pinker, S. 2002. *The Blank State*. Penguin.
10. Plotkin, H. 2002. *The Imagined World Made Real: Towards a Natural Science of Culture*. Penguin.
11. On television programme *The Cultured Ape*. Channel 4. Produced by Brian Leith, Scorer Associates.
12. de Waal, F. 2001. *The Ape and the Sushi Master*. Penguin.
13. Tomasello, M. 1999. *The Cultural Origins of Human Cognition*. Harvard University Press.
14. de Waal, F. 2001. *The Ape and the Sushi Master*. Penguin.
15. Tomasello, M. 1999. *The Cultural Origins of Human Cognition*. Harvard University Press.
16. Tiger, L. and Fox, R. 1971. *The Imperial Animal*. Transaction.
17. Rizzolatti, G., personal communication.
18. Rizzolatti, G. and Arbib, M.A. 1998. Language within our grasp. *Trends in Neurosciences* 21:188–94.
19. Iacobini, M., Koski, L.M., Brass, M., Bekkering, H., Woods, R.P., Dubeau, M.-C., Mazziotta, J.C. and Rizzolatti, G. 2001. Reafferent copies of imitated actions in the right superior temporal cortex. *Proceedings of the National Academy of Sciences* 98:13995–9.
20. Kohler, E., Keysers, C., Umilta, M.A., Fogassi, L., Gallese, V. and Rizzolatti, G. 2002. Hearing sounds, understanding actions: action representation in auditory mirror neurons. *Science* 297:846–8.
21. Lai, C.S., Fisher, S.E. *et al.* 2001. A forkhead-domain gene is mutated in a severe speech and language disorder. *Nature* 413: 519–23.
22. Enard, W., Przeworski, M., Fisher, S.E., Lai, C.S.L., Wiebe, V., Kitano, T., Monaco, A.P. and Paabo, S. 2002. Molecular evolution of FOXP2, a gene involved in speech and language. *Nature* 418:869–72.
23. Iacoboni, M., Woods, R.P., Brass, M., Bekkering, H., Mazziotta, J.C. and Rizzolatti, G. 1999. Cortical mechanisms of human imitation, *Science* 286:2526–8.

24. Cantalupo, C. and Hopkins, W.D. 2001. Asymmetric Broca's area in great apes. *Nature* 414:505.

25. Newman, A.J., Bavelier, D., Corina, D., Jezzard, P. and Neville, H.J. 2002. A critical period for right hemisphere recruitment in American Sign Language processing. *Nature Neuroscience* 5:76–80.

26. Dunbar, R. 1996. *Gossip, Grooming and the Evolution of Language.* Faber & Faber.

27. Walker, A. and Shipman, P. 1996. *The Wisdom of Bones.* Weidenfeld & Nicolson.

28. Tattersall, I. Email correspondence.

29. Wilson, F.R. 1998. *The Hand.* Pantheon.

30. Calvin, W.H. and Bickerton, D. 2001. *Lingua ex Machina.* MIT Press.

31. Stokoe, W.C. 2001. *Language in Hand: Why Sign Came before Speech.* Gallaudet University Press.

32. Durham, W.H., Boyd, R. and Richerson, P.J. 1997. Models and forces of cultural evolution. In *Human by Nature* (ed. Weingert, P., Mitchell, S.D., Richerson, P.J. and Maasen, S.). Lawrence Erlbaum Associates.

33. Deacon, T. 1997. *The Symbolic Species: the Co-evolution of Language and the Human Brain.* Penguin.

34. Blackmore, S. 1999. *The Meme Machine.* Oxford University Press.

35. Cronk, L. 1999. *That Complex Whole: Culture and the Evolution of Human Behavior.* Westview Press.

36. Pitts, M. and Roberts, M. 1997. *Fairweather Eden.* Century.

37. Kohn, M. 1999. *As We Know It: Coming to Terms with an Evolved Mind.* Granta.

38. Low, B.S. 2000. *Why Sex Matters: a Darwinian Look at Human Behavior.* Princeton University Press.

39. Dunbar, R., Knight, C. and Power, C. 1999. *The Evolution of Culture.* Edinburgh University Press.

40. Whiten, A. and Byrne, R.W. (eds). 1997. *Machiavellian Intelligence II.* Cambridge University Press.

41. Wright, R. 2000. *Nonzero: History, Evolution and Human Cooperation.* Random House.

42. Ridley, M. 1996. *The Origins of Virtue.* Penguin.

43. Ofek, H. 2001. *Second Nature.* Cambridge University Press.

44. Tattersall, I. 1998. *Becoming Human.* Harcourt Brace.

45. Wright, R. 2000. *Nonzero: History, Evolution and Human Cooperation.* Random House.

46. Ridley, M. 1996. *The Origins of Virtue*. Penguin.
47. Neville-Sington, P. and Sington, D. 1993. *Paradise Dreamed: How Utopian Thinkers Have Changed the World*. Bloomsbury.
48. Milne, H. 1986. *Bhagwan: The God That Failed*. Caliban Books.

CHAPTER 9: THE SEVEN MEANINGS OF GENE

1. Dennett, D. *Darwin's Dangerous Idea*. Penguin.
2. De Vries, H. 1900. Sur la loi de disjonction des hybrides. *Comptes Rendus de l'Académie des Sciences* (Paris) 130:845–7.
3. Henig, R.M. 2000. *A Monk and Two Peas*. Weidenfeld & Nicolson.
4. Tudge, C. 2001. *In Mendel's Footnotes*. Vintage; Orel, V. 1996. *Gregor Mendel: the First Geneticist*. Oxford University Press.
5. Watson, J.D. and Crick, F.H.C. 1953. Molecular structure of nucleic acid: a structure for deoxyribonucleic acid. *Nature* 171:737. Watson, J. with Barry, A. 2003. *DNA: The secret of Life*. Knopf.
6. Ptashne, M. and Gann, A. 2002. *Genes and Signals*. Cold Spring Harbor Press.
7. Midgley, M. 1979. Gene juggling. *Philosophy* 54:439–58.
8. Canning, C. and Lovell-Badge, R. Sry and sex determination: how lazy can it be? *Trends in Genetics* 18:111–13.
9. Randolph Nesse, personal communication.
10. Chagnon, N. 1992. *Yanomamo: the Last Days of Eden*. Harcourt Brace.
11. Miller, G. 2000. *The Mating Mind*. Doubleday.
12. Wilson, E.O. 1994. *Naturalist*. Island Press.
13. Wilson, E.O. 1975. *Sociobiology*. Harvard University Press.
14. Segerstrale, U. 2000. *Defenders of the Truth*. Oxford University Press.
15. Anthony Leeds, Barbara Beckwith, Chuck Madansky, David Culver, Elizabeth Allen, Herb Schreier, Hiroshi Inouye, Jon Beckwith, Larry Miller, Margaret Duncan, Miriam Rosenthal, Reed Pyeritz, Richard C. Lewontin, Ruth Hubbard, Steven Chorover and Stephen Gould 1975. Letter to the *New York Review of Books*. 13 November 1975.
16. Segerstrale, U. 2000. *Defenders of the Truth*. Oxford University Press.
17. Lewontin, R. 1993. *The Doctrine of DNA: Biology of Ideology*. Penguin.
18. Tooby, J. and Cosmides, L. 1992. The psychological foundations of

culture. In *The Adapted Mind* (ed. Barkow, J.H., Cosmides, L. and Tooby, J.). Oxford University Press.
19. Ibid
20. Daly, M. and Wilson, M. 1988. *Homicide.* Aldine.
21. Hrdy, S. 2000. *Mother Nature.* Ballantine Books.
22. Durkheim, E. 1895. *The Rules of the Sociological Method* (1962 edition). Free Press.

CHAPTER 10: A BUDGET OF PARADOXICAL MORALS

1. Wolfe, T. 2000. *Hooking Up.* Picador.
2. Ellis, B.J. and Garber, J. 2000. Psychosocial antecedents of variation in girls' pubertal timing: maternal depression, stepfather presence, and marital and family stress. *Child Development* 71:485–501.
3. Harris, J.R. 1998. *The Nurture Assumption.* Bloomsbury.
4. Harris, J.R. 1995. Where is the child's environment? A group socialisation theory of development. *Psychological Review.* 102:458–9.
5. Wills, J.E. 2001. *1688: a Global History.* Granta.
6. But then other studies reveal some rather dramatic negative correlations between parents and children, too: where the effect of the parent is to drive the child to do the opposite. The children of hippies become investment bankers.
7. Lytton, H. 2000. Towards a model of family-environmental and child-biological influences on development. *Developmental Review* 20:150–79.
8. Mednick, S.A., Gabrielli, W.F. and Hutchings, B. 1984. Genetic influences in criminal convictions: evidence from an adoption cohort. *Science* 224:891–4.
9. Scarr, S. 1996. How people make their own environments: implications for parents and policy makers. *Psychology Public Policy and Law* 2:204–28.
10. Collins, W.A., Maccoby, E.E., Steinberg, L., Hetherington, E.M. and Bernstein, M.H. 2000. The case for nature and nurture. *American Psychologist* 55:218–32.
11. Bennett, A.J., Lesch, K.P., Heils, A., Long, J.C., Lorenz, J.G., Shoaf, S.E., Champoux, M., Suomi, S.J. Linnoila, M.V. and Higley, J.D. 2002. Early experience and serotonin transporter gene variation interact to influence primate CNS function. *Molecular Psychiatry* 7:118–22.

12. Smith, A. 1776. *The Wealth of Nations*. London.
13. Ibid.
14. Durkheim, E. (1933). *The Division of Labor in Society*. Free Press.
15. Buss, D.M. 1994. *The Evolution of Desire*. Basic Books.
16. Lewontin, R.C. 1972. The apportionment of human diversity. *Evolutionary Biology* 6:381–98.
17. Kurzban, R., Tooby, J. and Cosmides, L. 2001. Can race be erased? Coalitional computation and social categorization. *Proceedings of the National Academy of Sciences* 98:15387–92.
18. Caspi, A., McClay, J., Moffitt, T., Mill, J., Martin, J., Craig, I.W., Taylor, A. and Poulton, R. 2002. Role of genotype in the cycle of violence in maltreated children. *Science* 297:851–4. Also, Terrie Moffitt and Avshalom Caspi, email correspondence. See also the Nuffield Council on Bioethics report (2002). *Genetics and Behavior: the Ethical Context*. Incidentally, as Judith Rich Harris pointed out to me, the correlation between parental abuse and antisocial behaviour in the Dunedin study cannot be assumed to be causal. It may be that another undiscovered gene affects both: a long history of fallacious assumption teaches us to be cautious before assuming that parents cause effects in children by their actions.
19. James, W. 1884. The dilemma of determinism. In *The Writings of William James* (ed. McDermott, J.J.). University of Chicago Press.
20. Walter, H. 2001. *Neurophilosophy of Free Will*. MIT Press.
21. Quoted by Walter, H. 2001. *Neurophilosophy of Free Will*. MIT Press.
22. *Hamlet*, Act 5, scene 2.
23. Freeman, W.J. 1999. *How Brains Make up Their Minds*. Weidenfeld & Nicolson.
24. Francis Crick, interview.
25. Tim Tully, interview.

EPILOGUE: *HOMO STRAMINEUS* – THE STRAW MAN

1. James, W. 1906. The moral equivalent of war. Address to Stanford University. Printed as Lecture 11 in *Memories and Studies*. Longman Green & Co. (1911):267–96.
2. Bouchard, T. 1999. Genes, environment and personality. In *The Nature–*

Nurture Debate: the Essential Readings (ed. Ceci, S.J and Williams, W.M.). Blackwell.

3. Wilson, E.O. 1978. *On Human Nature.* Harvard University Press.
4. Dawkins, R. 1981. Selfish genes in race or politics. *Nature* 289:528.
5. Pinker, S. 1997. *How the Mind Works.* Norton.
6. Rose, S. 1997. *Lifelines.* Penguin.
7. Gould, S.J. 1978. *Ever Since Darwin.* Burnett Books.
8. Lewontin, R. 1993. *The Doctrine of DNA: Biology as Ideology.* Penguin.
9. Tim Tully, interview.

INDEX